STATISTICAL THEORY AND RANDOM MATRICES

MONOGRAPHS AND TEXTBOOKS IN PURE AND APPLIED MATHEMATICS

1. *K. Yano,* Integral Formulas in Riemannian Geometry (1970) *(out of print)*
2. *S. Kobayashi,* Hyperbolic Manifolds and Holomorphic Mappings (1970) *(out of print)*
3. *V. S. Vladimirov,* Equations of Mathematical Physics (A. Jeffrey, editor; A. Littlewood, translator) (1970) *(out of print)*
4. *B. N. Pshenichnyi,* Necessary Conditions for an Extremum (L. Neustadt, translation editor; K. Makowski, translator) (1971)
5. *L. Narici, E. Beckenstein, and G. Bachman,* Functional Analysis and Valuation Theory (1971)
6. *D. S. Passman,* Infinite Group Rings (1971)
7. *L. Dornhoff,* Group Representation Theory (in two parts). Part A: Ordinary Representation Theory. Part B: Modular Representation Theory (1971, 1972)
8. *W. Boothby and G. L. Weiss (eds.),* Symmetric Spaces: Short Courses Presented at Washington University (1972)
9. *Y. Matsushima,* Differentiable Manifolds (E. T. Kobayashi, translator) (1972)
10. *L. E. Ward, Jr.,* Topology: An Outline for a First Course (1972) *(out of print)*
11. *A. Babakhanian,* Cohomological Methods in Group Theory (1972)
12. *R. Gilmer,* Multiplicative Ideal Theory (1972)
13. *J. Yeh,* Stochastic Processes and the Wiener Integral (1973) *(out of print)*
14. *J. Barros-Neto,* Introduction to the Theory of Distributions (1973) *(out of print)*
15. *R. Larsen,* Functional Analysis: An Introduction (1973) *(out of print)*
16. *K. Yano and S. Ishihara,* Tangent and Cotangent Bundles: Differential Geometry (1973) *(out of print)*
17. *C. Procesi,* Rings with Polynomial Identities (1973)
18. *R. Hermann,* Geometry, Physics, and Systems (1973)
19. *N. R. Wallach,* Harmonic Analysis on Homogeneous Spaces (1973) *(out of print)*
20. *J. Dieudonné,* Introduction to the Theory of Formal Groups (1973)
21. *I. Vaisman,* Cohomology and Differential Forms (1973)
22. *B.-Y. Chen,* Geometry of Submanifolds (1973)
23. *M. Marcus,* Finite Dimensional Multilinear Algebra (in two parts) (1973, 1975)
24. *R. Larsen,* Banach Algebras: An Introduction (1973)
25. *R. O. Kujala and A. L. Vitter (eds),* Value Distribution Theory: Part A; Part B. Deficit and Bezout Estimates by Wilhelm Stoll (1973)
26. *K. B. Stolarsky,* Algebraic Numbers and Diophantine Approximation (1974)
27. *A. R. Magid,* The Separable Galois Theory of Commutative Rings (1974)
28. *B. R. McDonald,* Finite Rings with Identity (1974)
29. *J. Satake,* Linear Algebra (S. Koh, T. Akiba, and S. Ihara, translators) (1975)

30. *J. S. Golan*, Localization of Noncommutative Rings (1975)
31. *G. Klambauer*, Mathematical Analysis (1975)
32. *M. K. Agoston*, Algebraic Topology: A First Course (1976)
33. *K. R. Goodearl*, Ring Theory: Nonsingular Rings and Modules (1976)
34. *L. E. Mansfield*, Linear Algebra with Geometric Applications: Selected Topics (1976)
35. *N. J. Pullman*, Matrix Theory and Its Applications (1976)
36. *B. R. McDonald*, Geometric Algebra Over Local Rings (1976)
37. *C. W. Groetsch*, Generalized Inverses of Linear Operators: Representation and Approximation (1977)
38. *J. E. Kuczkowski and J. L. Gersting*, Abstract Algebra: A First Look (1977)
39. *C. O. Christenson and W. L. Voxman*, Aspects of Topology (1977)
40. *M. Nagata*, Field Theory (1977)
41. *R. L. Long*, Algebraic Number Theory (1977)
42. *W. F. Pfeffer*, Integrals and Measures (1977)
43. *R. L. Wheeden and A. Zygmund*, Measure and Integral: An Introduction to Real Analysis (1977)
44. *J. H. Curtiss*, Introduction to Functions of a Complex Variable (1978)
45. *K. Hrbacek and T. Jech*, Introduction to Set Theory (1978) *(out of print)*
46. *W. S. Massey*, Homology and Cohomology Theory (1978)
47. *M. Marcus*, Introduction to Modern Algebra (1978)
48. *E. C. Young*, Vector and Tensor Analysis (1978)
49. *S. B. Nadler, Jr.*, Hyperspaces of Sets (1978)
50. *S. K. Sehgal*, Topics in Group Rings (1978)
51. *A. C. M. van Rooij*, Non-Archimedean Functional Analysis (1978)
52. *L. Corwin and R. Szczarba*, Calculus in Vector Spaces (1979)
53. *C. Sadosky*, Interpolation of Operators and Singular Integrals: An Introduction to Harmonic Analysis (1979)
54. *J. Cronin*, Differential Equations: Introduction and Quantitative Theory (1980)
55. *C. W. Groetsch*, Elements of Applicable Functional Analysis (1980)
56. *I. Vaisman*, Foundations of Three-Dimensional Euclidean Geometry (1980)
57. *H. I. Freedman*, Deterministic Mathematical Models in Population Ecology (1980)
58. *S. B. Chae*, Lebesgue Integration (1980)
59. *C. S. Rees, S. M. Shah, and Č. V. Stanojević*, Theory and Applications of Fourier Analysis (1981)
60. *L. Nachbin*, Introduction to Functional Analysis: Banach Spaces and Differential Calculus (R. M. Aron, translator) (1981)
61. *G. Orzech and M. Orzech*, Plane Algebraic Curves: An Introduction Via Valuations (1981)
62. *R. Johnsonbaugh and W. E. Pfaffenberger*, Foundations of Mathematical Analysis (1981)

63. *W. L. Voxman and R. H. Goetschel,* Advanced Calculus: An Introduction to Modern Analysis (1981)
64. *L. J. Corwin and R. H. Szczarba,* Multivariable Calculus (1982)
65. *V. I. Istrătescu,* Introduction to Linear Operator Theory (1981)
66. *R. D. Järvinen,* Finite and Infinite Dimensional Linear Spaces: A Comparative Study in Algebraic and Analytic Settings (1981)
67. *J. K. Beem and P. E. Ehrlich,* Global Lorentzian Geometry (1981)
68. *D. L. Armacost,* The Structure of Locally Compact Abelian Groups (1981)
69. *J. W. Brewer and M. K. Smith, eds.,* Emmy Noether: A Tribute to Her and Work (1981)
70. *K. H. Kim,* Boolean Matrix Theory and Applications (1982)
71. *T. W. Wieting,* The Mathematical Theory of Chromatic Plane Ornaments (1982)
72. *D. B. Gauld,* Differential Topology: An Introduction (1982)
73. *R. L. Faber,* Foundations of Euclidean and Non-Euclidean Geometry (1983)
74. *M. Carmeli,* Statistical Theory and Random Matrices (1983)

Other Volumes in Preparation

STATISTICAL THEORY AND RANDOM MATRICES

Moshe Carmeli

Center for Theoretical Physics
Ben Gurion University of the Negev
Beer Sheva
Israel

MARCEL DEKKER, INC. New York and Basel

Library of Congress Cataloging in Publication Data

Carmeli, Moshe,
Statistical theory and random matrices.

(Monographs and textbooks in pure and applied mathematics ; v. 74)
Bibliography: p.
Includes index.
1. Energy levels (Quantum mechanics)--Statistical methods. 2. Matrices. I. Title. II. Series.
QC174.45.C37 1983 530.1'43 82-22031
ISBN 0-8247-1779-1

MARCEL DEKKER, INC.
270 Madison Avenue, New York, New York 10016

Current printing (last digit):
10 9 8 7 6 5 4 3 2 1

PRINTED IN THE UNITED STATES OF AMERICA

To

Eugene P. Wigner

PREFACE

The question whether the highly excited states of a physical system may be understood by assuming no structure to the system, and that no quantum number other than the spin and the parity remains good, leads to the *statistical theory of energy levels* and its relation to *random matrices*. Such a statistical theory is designed to describe the general appearance and the degree of irregularity of the level structure that occurs in a complex physical system, which is otherwise too complicated to be understood in detail, rather than to predict the detailed sequence of the energy levels in any particular nucleus or atom.

The standard type of statistical mechanics is clearly inadequate for the discussion of energy levels since statements on the *fine detail* of the energy level structure cannot be made in terms of an ensemble of states. What is required is a *different* kind of statistical mechanics in which one renounces the exact knowledge *not* of the state of the system but of the *nature* of the system itself.

The problem then is to define in a mathematically precise way an ensemble of systems in which all possible laws of interactions are equally probable. The idea of a statistical mechanics of nuclei, which is based on an ensemble of systems, is due to Wigner and to von Neumann. This book summarizes the fundamentals of this theory.

After introducing the basic concepts of the statistical theory of energy levels and their relations to random matrices in Chapters 1 and 2, we discuss the symmetry properties of physical systems in Chapter 3. Different kinds of ensembles are subsequently introduced and discussed. This includes the Gaussian and the orthogonal ensembles which are discussed in Chapter 4, followed by the unitary ensemble discussed in Chapter 5.

In Chapter 6 the problem of eigenvalue-eigenvector distributions of the Gaussian ensemble is discussed, while Chapter 7 deals with the distribution of the widths. We then discuss the symplectic group and its relation to quaternions in Chapter 8. A detailed discussion on the Gaussian ensemble is subsequently given in Chapter 9. Chapter 10 is devoted to the summary, whereas Appendices A and B include a detailed discussion on the statistical aspects of multivariate distributions, and Appendix C on the ergodic properties of random matrices.

I am indebted to Professor P. R. Krishnaiah for stimulating my interest in the statistical theory of multivariate distributions of random matrices. I am also most thankful to Professor E. P. Wigner for emphasizing the uniqueness of the statistical theory of energy levels and for his comments on the use of complex Wishartian ensemble (see Chapter 10). Also, I am indebted to Mrs. Sara Corrogosky for her technical assistance in preparing and typing the manuscript of the book, and to Professor J. B. French and the American Physical Society for their permission to reproduce Figures 5.1-5.5, 6.1-6.3, and 7.1.

Part of this book was written while the author was a Visiting Professor and Member of the Institute for Theoretical Physics, State University of New York at Stony Brook. I am indebted to Professor Chen Ning Yang for his kind hospitality at Stony Brook in 1977-78 and 1981. Finally, I am most grateful to my wife Elisheva for her moral encouragement and technical help without which the book would probably have never been written.

Moshe Carmeli

CONTENTS

PREFACE		v
Chapter 1	INTRODUCTION	1
1.1	Complex systems	1
1.2	Statistical Theory of Levels	2
1.3	Comparison with Statistical Mechanics	3
1.4	Examples	4
1.5	Distribution of the Widths	7
1.6	Preview	8
Chapter 2	HIGHLY EXCITED SYSTEMS	9
2.1	One-Dimensional Systems	9
2.2	Sequences of Numbers	10
2.3	Quantal Spectra	12
Chapter 3	SYMMETRY PROPERTIES OF PHYSICAL SYSTEMS	15
3.1	Symmetry Properties	15
3.2	Time-Reversal	16
3.3	The Hamiltonian Matrix	18
3.4	Relation to Kramers' Degeneracy	19
3.5	Quaternion Structure	20
3.6	The Threefold Way of Invariance	21

Chapter 4 GAUSSIAN AND ORTHOGONAL ENSEMBLES 23
4.1 The Gaussian Ensemble 23
4.2 The Orthogonal Ensemble 24
4.3 The Semicircle Distribution 25
4.4 The Wishart Ensemble 26
4.5 Invariance and Equal Probability 27

Chapter 5 UNITARY ENSEMBLE 29
5.1 Systems without Time-Reversal Symmetry 29
5.2 Level Repulsion in Spectra 31
5.3 Repulsion in Low Energy Spectra 39

Chapter 6 EIGENVALUE-EIGENVECTOR DISTRIBUTIONS OF THE GAUSSIAN ENSEMBLE 43
6.1 The Gaussian Ensemble 43
6.2 The Repulsion of Levels 46
6.3 Wigner's Conjecture 47
6.4 Mean Level Density 49
6.5 Remark 50
6.6 The Spectral Rigidity 50

Chapter 7 DISTRIBUTION OF THE WIDTHS 57
7.1 Widths of Energy Levels 57
7.2 The Porter-Thomas Distribution 58
7.3 Strength Fluctuations and Collective Behavior 60
7.4 The Porter-Thomas Distribution and Experimental Data 62

Chapter 8 SYMPLECTIC GROUP AND QUATERNIONS 67
8.1 The Symplectic Group 67
8.2 Quaternions 68
8.3 Matrices and Quaternions 69
8.4 Quaternion Algebra 70
8.5 Symplectic Ensemble 72

Chapter 9 MORE ON THE GAUSSIAN ENSEMBLE 75
9.1 The Gaussian Ensemble 75
9.2 The Distribution Function 76
9.3 The Jacobian of the Transformation 78
9.4 Specification of the Distribution Function 79
9.5 The Unitary Ensemble Case 80

Chapter 10 SUMMARY 83
10.1 Random Matrices and Energy Levels 83
10.2 Role of the Hamiltonian 84
10.3 The Wishart Distribution 85
10.4 Recent Studies 85
10.5 Finite and Infinite Matrices 86

Appendix A MULTIVARIATE DISTRIBUTIONS 87
A1. Distributions 87
A2. Preliminaries 88
A3. Distributions of Some Random Matrices 94
A4. Marginal Distributions of Few Roots 100
A5. Moments of the Elementary Symmetric Functions 104
A6. Distributions of the Ratios of the Roots 108
A7. Distributions of the Likelihood Ratio Test Statistics 109

Appendix B APPLICATIONS OF MULTIVARIATE DISTRIBUTIONS 115
B1. Preliminaries 115
B2. Testing Hypotheses 116
B3. Discrimination between Multivariate Normal Population 122
B4. Alternative Procedures 123

Appendix C RANDOM MATRICES: ERGODIC PROPERTIES 129
C1. Preliminaries 130
C2. One-Point Measures 135
C3. Higher-Order Functions 139
C4. Behavior of the Strength Distribution 146
C5. Final Remarks 150

REFERENCES 151

BIBLIOGRAPHY 169

INDEX 179

STATISTICAL THEORY AND RANDOM MATRICES

CHAPTER 1
INTRODUCTION

1.1 COMPLEX SYSTEMS

Within the theory of quantum mechanics [see, for instance, Schiff (*1*)] the behavior of a physical system is determined by a state function Ψ. The state function is then a solution of the familiar Schrödinger equation,

$$H\Psi = E\Psi, \tag{1.1}$$

where H is the *Hamiltonian operator*, which is a *Hermitian operator*, and E is a constant which denotes the energy levels of the system.

Thus the energy levels are *characteristic values* (*eigenvalues*, or *roots*) of Hermitian operators. The *stationary states* of the system are the corresponding *characteristic vectors* (or *eigenfunctions*).

Although theoretical analyses have had impressive success, as was pointed out by Kisslinger and Sorensen (*2*), and by Baranger (*3*), in interpreting the detailed structure of the *low-lying* excited states of complex systems [here the word system is used for a physical quantum system that can be described by the Schrödinger equation; a system could be, for example, a complex nucleus, as is discussed in (*2*) and (*3*), or an atomic system], still, there must come a point beyond which such analyses of individual levels cannot usefully go [Dyson (*4*)]. For example, observations of levels of heavy nuclei in the neutron-capture region [Rosen, Desjardins, Rainwater, and Havens (*5*)] give exact information on the energy levels from number N to number $N + n$, where n is an integer of the order of 100 whereas N is an integer of the order of one million. It appears improbable that energy level assignments, based on various models, can ever be pushed as far as the millionth level.

1.2 STATISTICAL THEORY OF LEVELS

One is then led to ask whether the highly excited states may be understood from the opposite point of view, by assuming no structure for the system and that no quantum numbers other than spin and parity remain good. Such an inquiry leads to a *statistical theory of energy levels*.

Such a statistical theory is not supposed to predict the detailed sequence of energy levels in any one nucleus or atom, but is expected to describe the *general appearance* and the *degree of irregularity* of the level structure that is to occur in a complex system which is otherwise too complicated to be understood in detail.

As Dyson (*4*) has pointed out, in ordinary statistical mechanics a comparable renunciation of exact knowledge about the system is also made. By assuming that all states of a very large ensemble are equally probable, one obtains useful information about the overall behavior of a complex system when the observation of the state of the system in all its detail is impossible. This standard type of statistical mechanics is clearly inadequate for the discussion of energy levels. What one wishes is to make statements on the fine detail of the energy level structure, and such statements cannot be made in terms of an ensemble of states. What is required is a *different* kind of statistical mechanics in which one renounces exact knowledge *not* of the state of a system but of the *nature* of the system itself. One might picture a complex nucleus as a "black box" in which a large number of particles are interacting according to unknown laws. The problem then is to define in a mathematically precise way an ensemble of systems in which all possible laws of interaction are equally probable. The idea of a statistical mechanics of nuclei based on an ensemble of systems is due to Wigner.

1.3 COMPARISON WITH STATISTICAL MECHANICS

The difference between the usual statistical mechanics and the statistical theory of energy levels can also be seen, according to Wigner (*6*), as follows.

A system in quantum mechanics can be characterized by the Hamiltonian H, which is a self-adjoint linear operator in the infinite-dimension Hilbert space of functions Ψ. If one introduces a coordinate system in the Hilbert space, the Hamiltonian operator may then be looked at as a Hermitian matrix of infinitely many dimensions. Therefore, an ensemble of systems can be considered as an ensemble of Hermitian matrices. At this stage one might consider matrices of very high dimensionality rather than infinite matrices. However, the question arises as to what ensemble of such matrices one has to consider. Herein lies the difference between the ensembles of statistical mechanics and the ensemble of the statistical theory of energy levels.

In statistical mechanics one considers a system of particles with definite masses interacting among themselves by a given law. The state of such a system can be specified, in classical mechanics, by the generalized coordinates q_i and the generalized momenta p_i of the particles, where both q_i and p_i are functions of time. The physical quantities one is then interested in are the time averages of continuous functions f of the coordinates and momenta,

$$\lim_{T\to\infty}\left(\frac{1}{T}\int_t^{t+T} f(q_1(\tau), q_2(\tau), \ldots, p_1(\tau), p_2(\tau), \ldots)\, d\tau\right). \tag{1.2}$$

Using Newton's law of motion, one can, in principle, determine the coordinates and momenta as functions of time and their initial values [see, e.g., Goldstein (*7*)]. Hence the averaging process is an entirely *definite* one, and the result is a function only of the constants of motion, such as energy, but independent of other initial conditions. This result, except for rare cases, has long been proved and known by von Neumann and others [Koopman (*8*); Birkhoff (*9, 10*); von Neumann (*11*); Birkhoff and Koopman (*12*)].

The averaging process in the theory of random processes, on the other hand, is *not* defined. One again deals with a specific system with its proper, though in many cases unknown, Hamiltonian and pretends that one deals with a multitude of systems, all with their own Hamiltonians, and averages over the properties of these systems. Such a procedure can be meaningful only if it turns out that the properties in which one is interested are the same for the vast majority of the admissible Hamiltonians.

1.4 EXAMPLES

What are the admissible Hamiltonians, and what is the proper measure in the ensemble of these Hamiltonians? And suppose the ensemble of admissible Hamiltonians with a proper measure is given. Are the properties in which we are interested common for the vast majority of them?

Figures 1.1-1.3 illustrate the situation which leads to the idea of the statistical properties of the spectrum in the higher-energy region, as compared to low-energy region, where one desires to have a rather complete description of the stationary states and as complete a listing as possible of the exact values of the energy levels.

Figure 1.1 gives the energy levels of the nuclei beryllium, boron, and carbon (^{10}Be, ^{10}B, and ^{10}C). The diagram shows the eight lowest energy levels of ^{10}B and the lowest two energy levels of ^{10}Be and ^{10}C. It gives the position of these energy levels, their total angular momenta J, and parities T [Ajzenberg and Lauritsen (*13*)].

Of much interest, but not shown in the diagram, are the transition probabilities between these levels. Such transition probabilities can be calculated if the characteristic functions associated with the characteristic values are known. Conversely, agreement between the

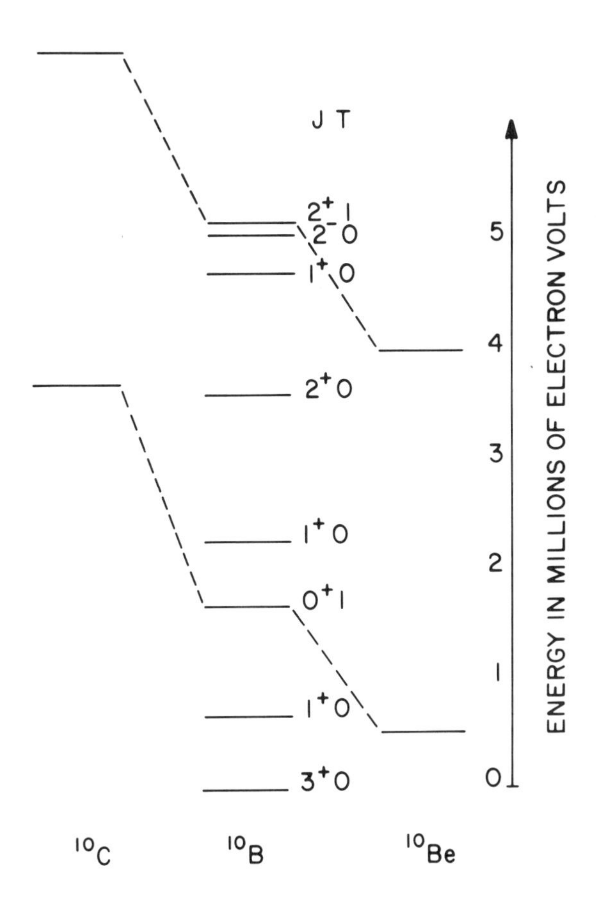

Figure 1.1 Energy levels of the nuclei ^{10}Be, ^{10}B, and ^{10}C [Ajzenberg and Lauritsen (*13*)].

observed transition probabilities and the calculated values of these quantities gives an indication of the accuracy of the calculated characteristic function [Wigner (*14*)].

Figure 1.2 gives the energy levels of ^{180}Hf. This nucleus has a rotational band [Mihelich, Scharff-Goldhaber, and McKeown (*15*)]. The angular momenta of the states shown are $J = 0, 2, 4, 6, 8$ in units of $h/2\pi$, where h is Planck's constant. The energy levels of these states are proportional to $J(J + 1)$, where J is the angular momentum quantum number.

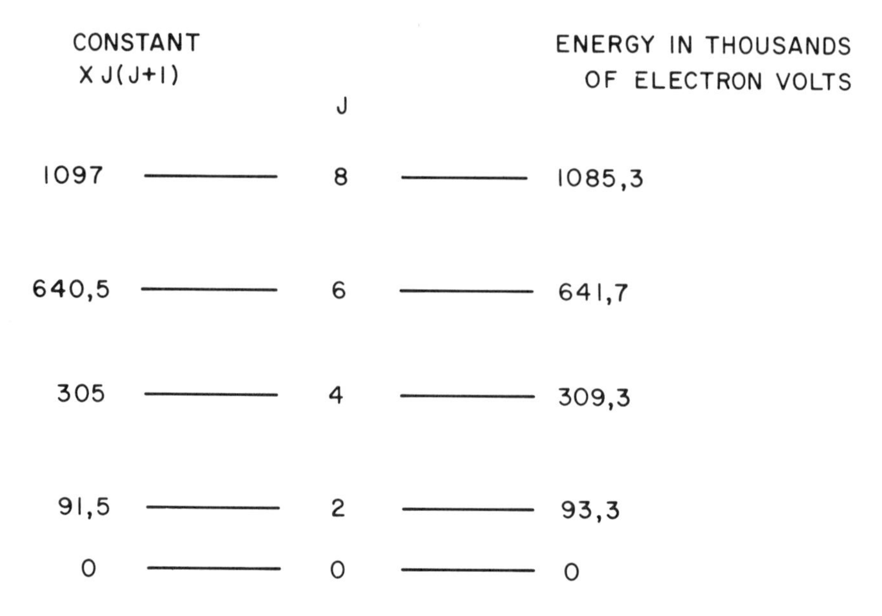

Figure 1.2 Energy levels of ^{180}Hf [Mihelich *et al.* (*15*)].

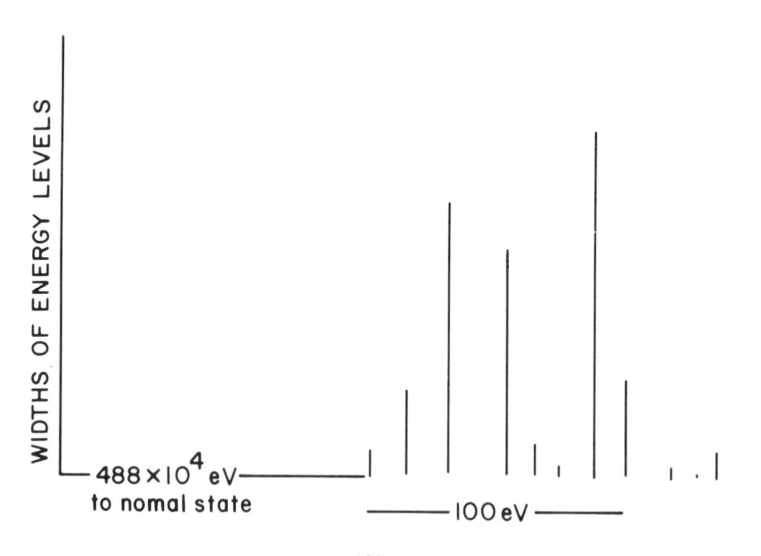

Figure 1.3 Energy levels of ^{239}U [Hughes and Harvey (*16*)].

Figure 1.3 gives the energy levels of ^{239}U, the angular momentum quantum number of which is one-half [Hughes and Harvey (*16*)]. The diagram extends over 200 eV and its lowest point is about 4.88 MeV over the lowest energy level. It is of little interest and is almost impossible to calculate the exact position of these energy levels. The reason one knows their position with the accuracy shown in the diagram is that the addition of a low-energy neutron to a ^{238}U nucleus gives the ^{239}U nucleus with an energy of about 4.88 MeV.

The last diagram gives an example of energy levels in a region where one will be interested mostly in the statistical statements such as the density of the energy levels, their average width (i.e. the square of the wave function at the nuclear boundary), etc. Furthermore, one is also interested in the probability for certain spacings, including the question of whether the levels are, on the whole, equidistant or distributed according to a certain probability law.

1.5 DISTRIBUTION OF THE WIDTHS

In addition to the average width of the levels, one is interested in the distribution of the widths, i.e., the fraction of levels the widths of which are in unit interval at a certain width.

From the point of view of mathematics, the statistical questions are far more interesting than the question of the exact properties of the low-lying energy levels. This is so since it is likely that the statistical properties of a large class of real symmetric operators are in many respects identical. They should depend then on only a few parameters which are characteristic of the problem.

For example, one may confine one's attention to the class of real symmetric operators since the energy operators are not only Hermitian but are also real. This statement is a result of the time inversion symmetry of most physical systems. Notice, however, that not all physical systems have time symmetry. This was first pointed out as a result of experiments which show a violation of *CP* invariance in the decay $K_2{}^0 \rightarrow \pi^+ + \pi^-$ in 1964 by Christenson, Cronin, Fitch, and Turlay (*17*).

1.6 PREVIEW

In the following the physical aspects of the statistical theory of the energy levels of complex physical systems and their relation to the mathematical theory of random matrices, following Carmeli (*18*), are discussed.

After a preliminary introduction we summarize the symmetry properties of physical systems. Different kinds of ensembles are subsequently discussed. This includes the Gaussian, orthogonal, and unitary ensembles. The problem of eigenvalue-eigenvector distributions of the Gaussian ensemble is then discussed, followed by a discussion on the distribution of the widths. Finally, we discuss the symplectic group and quaternions, and the Gaussian ensemble in detail. In the appendices a review of some of the developments of complex multivariate distributions is given; also are given the ergodic properties of random matrices.

In the next chapter, we will discuss the spectroscopy of highly excited complex systems.

CHAPTER 2
HIGHLY EXCITED SYSTEMS

2.1 ONE-DIMENSIONAL SYSTEMS

What guidance has one in order to phrase the questions to be answered in the spectroscopy of highly excited complex systems? To start with, one recalls the energy levels comprising the spectra of a few well-known one-dimensional quantum mechanical problems [Porter (*19*)]:

$$\begin{aligned}
&\text{Harmonic oscillator}: && E_n \cong n + 1/2, \\
&\text{Infinite square well}: && E_n \cong (n+1)^2, \\
&\text{Hydrogen atom}: && E_n \cong -\frac{1}{(n+1)^2}, \\
& && \text{and continuum for } E > 0,
\end{aligned}$$

where $n = 0, 1, 2, \ldots$. For each of the three potentials there is another constant of the motion, the parity, which acts as an additional quantum number.

In the one-dimensional hydrogen atom case all energy levels, except the ground state, are doubly parity degenerate (i.e., the even solutions which are rejected by boundary condition requirements in the three-dimensional hydrogen s state are present in the one-dimensional case). Degeneracy is a statement concerning energy level spacings. Accordingly, in the case of the one-dimensional hydrogen atom the spacing between corresponding levels of even and odd parity is zero.

If one ignores the question of degeneracy associated with parity labeling and also ignores the existence of a continuum for positive energies for the one-dimensional hydrogen atom, one sees that the E_n form a countable sequence of numbers. Thus one might learn something about the positions of energy levels of highly excited states by examining the mathematics of number sequences.

2.2 SEQUENCES OF NUMBERS

What sequences of numbers does one know in mathematics? Perhaps the best-known sequence of numbers which is not trivial is the sequence of prime numbers. It is well known that no simple formula for the nth prime number p_n exists, although there are formulas involving limiting operations.

The best-known result concerning the prime numbers is the so-called prime number theorem [van der Pol (*20*); van der Pol and Bremmer (*21*)]. It gives an answer to the question of how many primes $N(n)$ there are between the integers 1 and n for a large integer n. The result, for large n, is

$$N(n) = \int_0^n \frac{dt}{\log t} \cdot \tag{2.1}$$

In Fig. 2.1 this function is plotted as a function of n. [This figure appears in Porter (*19*). It is based on Lehmer's (*22*) tables.] The density of prime numbers $dN(n)/dn$, given by

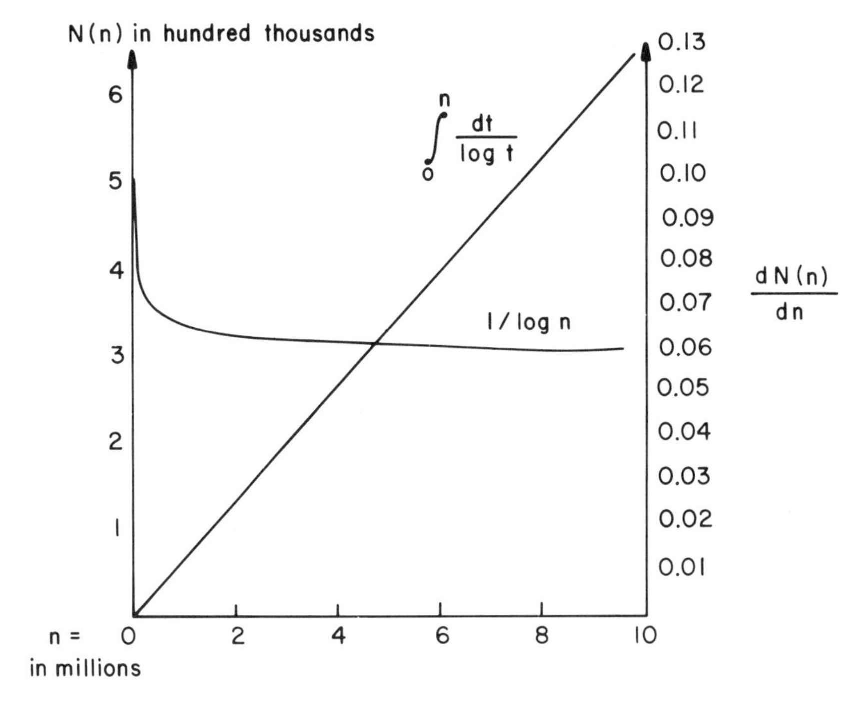

Figure 2.1 Cumulative distribution N(n) of prime numbers versus integers n, and the density dN/dn of the prime numbers [Porter (*19*), based on tables in Lehmer (*22*)].

$$\frac{dN(n)}{dn} = \frac{1}{\log n}, \tag{2.2}$$

is also given in this figure.

It will be noted that statistical concepts are already involved in the discussion of the prime numbers, such as the notion of the density of the primes. This is similar to the situation one finds in quantal spectra where, given a Hamiltonian, there is nothing random about the solution of the Schrödinger equation. On the other hand, there is apparently enough complexity in the prime number sequence that the density is not without what might be considered as fluctuations about the mean density.

What then would one mean by a completely random sequence? For an experimentalist this is a rather simple question since it is the same problem as the determination of the singles, doubles, etc., counting rate for a decaying radiative source. The major property of a decaying source is not that the decay occurs at random, but that the events are ordered in time.

It is well known that the relevant decay law for the differential probability $P^k(x)$, where $x = t/\tau$ is the time measured in units of the mean life τ of the source between two counts having k counts occur between them, is the Poisson distribution

$$P^k(x) = \frac{x^k}{k!} e^{-x}. \tag{2.3}$$

Hence one expects the nearest-neighbor spacing distribution for an ordered sequence of random numbers to be $P^0(x) = \exp(-x)$ and the next-nearest-neighbor spacing distribution to be $P^1(x) = x \exp(-x)$, etc., where $x = S/D$ is the spacing measured in units of the mean distance D between random numbers.

2.3 QUANTAL SPECTRA

Let us now return from the mathematical digression to the problem of quantal spectra. Most of quantum spectroscopy is based on the Schrödinger equation given in the introduction, where the operator (Hamiltonian) H is taken in the form

$$H(1, 2, \ldots, A) = \sum_{i=1}^{A} (T_i + V_i) + \sum_{i<j=1}^{A} V_{ij} \tag{2.4}$$

for a system of A particles when one assumes that there is a two-body interaction potential. The Schrodinger equation is then solved for the eigenvalues E (which give the spacing S) and the wave functions Ψ. Once the wave functions are found one can then calculate expectation values of operators such as magnetic dipoles, electric quadrupoles, and so on. One can also calculate widths of energy levels. The situation is summarized in Fig. 2.2.

The case for which the discrete levels under consideration are unstable to particle emission is somewhat more complicated in practice since there is often a background amplitude that cannot be neglected. This is so since the scheme presented in Fig. 2.2 is

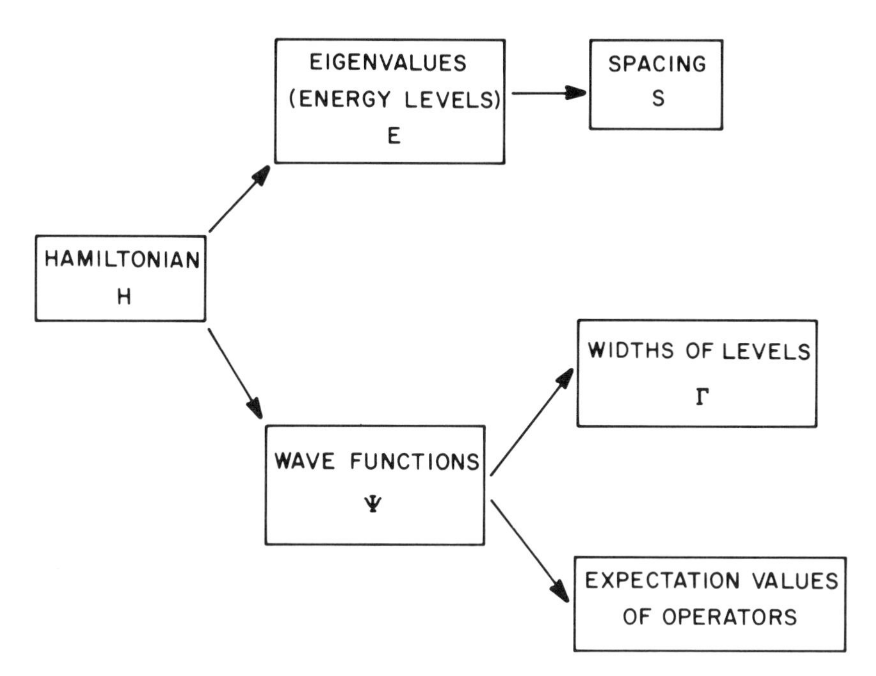

Figure 2.2 Typical pattern of a theoretical calculation of energy levels [Porter (*19*)].

typically used where the potential scattering is negligible, namely, for low excited states.

In the next chapter, the symmetry properties of physical systems will be discussed.

CHAPTER 3
SYMMETRY PROPERTIES OF PHYSICAL SYSTEMS

3.1 SYMMETRY PROPERTIES

The symmetry properties of any physical system are an expression of the Hamiltonian of the system [Porter (*19*); also e.g. Landau and Lifshitz (*23*)]. Thus, for example, if the Hamiltonian H is independent of time, then the energy of the system is conserved. Hence energy conservation is associated with invariance of the Hamiltonian under time translation.

For an isolated system of particles with interactions depending only on the distances between the particles, or invariant combinations of spin and interparticle coordinate vectors, translations along the spatial axes of a chosen coordinate system then leave the Hamiltonian unchanged, thus leading to the law of conservation of total linear momentum of the system.

The conservation of angular momentum can be related to rotations in an appropriate way. One can subtract from the total energy the kinetic energy associated with the total momentum. This leads to the notion of internal energy of the isolated system which is also conserved.

The internal energy of the complex system is of great interest for the statistical theory of energy levels since the latter deals with spectra arising from changes in the internal state of excitation of the system. Spatial inversion transformation (improper transformation) leads, in an odd number of dimensions, to changing the sign of every coordinate q_i and momentum p_i in the Hamiltonian.

If the Hamiltonian remains unchanged by this transformation (in technical language this means if the parity transformation operator commutes with the Hamiltonian, as is the case for gravitational, electromagnetic, and strong interactions), then the parity of the system is also conserved. Although the inversion transformation operator does not commute with the translation operators, there are simultaneous eigenstates of the translation operator and the parity operator so that the energy, total linear momentum, and parity become simultaneously constants of the motion.

3.2 TIME-REVERSAL

We will now come back to the reversal of the direction of time. A measurable physical quantity is invariant under a canonical transformation. For example, the quantity $|(\Psi, \Phi)|^2$, where the bracket describes a scalar product between the two-state functions Ψ and Φ, is invariant under the canonical transformations $\Psi \rightarrow A\Psi$, $\Phi \rightarrow A\Phi$. This means [Wigner (*24*); Rosenzweig (*25*); Messiah (*26*); Wick (*27*)]:

$$|(\Psi, \Phi)|^2 = |(A\Psi, A\Phi)|^2. \tag{3.1}$$

To satisfy Eq. (3.1), the operator A should be *unitary*, i.e., $AA^\dagger = 1$. But A could be either *linear* or *antilinear*, depending on whether it satisfies the equation

$$A(a\Psi + b\Phi) = a(A\Psi) + b(A\Phi), \tag{3.2}$$

or

$$A(a\Psi + b\Phi) = \bar{a}(A\Psi) + \bar{b}(A\Phi), \tag{3.3}$$

where $\bar{a}$ and $\bar{b}$ are the complex conjugates of a and b. The two possibilities (3.2) and (3.3) are uniquely compatible with the physical requirement (3.1).

The symmetry operators corresponding to space-time translations and spatial inversions are linear in the coordinate representation, whereas the time inversion operator is antilinear. The last result can also be inferred by considering the time-dependent Schrödinger equation,

$$H\Psi = i\hbar \frac{\partial \Psi}{\partial t}. \tag{3.4}$$

If one expands the solution Ψ in terms of eigenstates Ψ_k,

$$\Psi(t) = \sum_k C_k e^{-iE_k t/\hbar} \Psi_k, \tag{3.5}$$

the result of replacing t by $-t$ is equivalent to replacing $\exp(-iE_k t/\hbar)$ by its complex conjugate, thus showing that the operation of time inversion is related to complex conjugation, an antilinear operation.

This is also seen from the operator $i\hbar\,\partial/\partial t$ appearing on the right-hand side of the Schrödinger equation (3.4), where complex conjugation means time reversal. Time reversal does not change the coordinate q but reverses the direction of the momentum p since the latter involves a time rate. Denoting the time-reversal operator by T, one then has $TqT^{-1} = q$, but $TpT^{-1} = -p$. Applying these relations to the commutation relation $[q, p] = i\hbar$ leads to $TiT^{-1} = -i$, which shows again the antilinear effect of T.

It thus follows that T is antiunitary. However, if one defines a complex conjugation operator K such that $K\Psi = \bar{\Psi}$, then $K^2 = 1$ and the combination TK is unitary. T can then be written as $T = Uk$, where U is unitary. A simple calculation then shows that $T^2 = \pm 1$.

If the particles of the system do not have spin, then the unitary operator U can be chosen in the coordinate representation to be the identity operator. If the particles have spins, however, the choice of U is determined by the total angular momentum J, which satisfies $TJT^{-1} = -J$. The operator K gives the needed behavior for the orbital part of angular momenta in the coordinate representation. For one particle with spin, one can represent T as

$$T = e^{\frac{1}{2} i \pi \sigma_y} K, \tag{3.6}$$

where σ_y is one of Pauli's spin matrices. For a system having A particles, T can be represented as

$$\begin{aligned} T &= e^{\frac{1}{2} i \pi (\sigma_{1y} + \cdots + \sigma_{Ay})} K \\ &= e^{\frac{1}{2} i \pi S_y} K, \end{aligned} \tag{3.7}$$

which has the property $T^2 = 1$ for even A and $T^2 = -1$ for odd A.

3.3 THE HAMILTONIAN MATRIX

One thus has the set of commuting operators shown in Table III-1. Can any conclusion be reached about the structure of the Hamiltonian matrix from these general invariance properties? If the Hamiltonian has none of the symmetries mentioned above, then the Hamiltonian matrix must be complex Hermitian and hence is not real in general. The appropriate canonical transformation group is therefore the unitary group which preserves the Hermitian property of the Hamiltonian under a similarity transformation.

Suppose now that the Hamiltonian is time-reversal invariant. Then one faces two situations according to whether the total angular

Table III - 1

Invariance	Operator
Space translation	Total linear momentum
Time translation	Total energy
Space inversion	Parity
Time inversion	Time reversal
Space rotation	Total angular momentum

momentum of the system in units of ħ is integral or half-integral, thus depending upon whether $T^2 = +1$ or $T^2 = -1$. The $T^2 = 1$ case is easily handled and the Hamiltonian can be made real by an appropriate choice of basis. If $T^2 = -1$, again the Hamiltonian matrix can be made real provided that it is rotationally invariant.

Accordingly, the Hamiltonian matrix can be made real provided it is time-reversal invariant and either the system has integral spin or, if it has half-integral total angular momentum, the Hamiltonian is invariant under rotations. Thus one deals with real symmetric matrices and canonical transformations that correspond to a change from one basis in which the Hamiltonian is symmetric and real to another basis in which it remains the same. The aggregate of such transformations R provides a group of transformations, the *orthogonal group*,

$$RR^t = 1, \tag{3.8}$$

where R^t is the transpose of the matrix R.

3.4 RELATION TO KRAMER'S DEGENERACY

The situation for which the spin of the system is half-integral and the Hamiltonian is time-reversal invariant, but the total angular momentum is not conserved, such as that of an atom located in a multipole external crystalline electric field, is related to the well-known Kramers degeneracy [Kramers (*28*); Tinkham (*29*)], where Ψ and $T\Psi$ are orthogonal. Since both Ψ and $T\Psi$ satisfy the Schrödinger equation (H is time-reversal invariant) for the same energy, there is at least a doublet degeneracy in this case.

The Kramers degeneracy appears in the structure of the Hamiltonian matrix in an intrinsic way and may be seen as follows [Dyson (*4*); Wilson (*30*)]. Since $T^2 = -1$, one has $UKUK = -1$, or $UU = -1$. Since U is unitary, one sees that it must be skew-symmetric. Under a unitary basis transformation V, $T \to VTV^{-1}$, and hence $U \to VUV^t$. Under such a transformation the skew-symmetric unitary matrix U can be brought into a form $Z_{ij} = \delta_{i,j+1} - \delta_{i,j-1}$, i.e., a banded diagonal matrix with $+1$ in the subdiagonal, -1 in the superdiagonal [Hua (*31*)].

Further canonical unitary transformations S are possible if they commute with the time-reversal operator $T = ZK$. Hence one requires for S

$$ZKS - SZK = 0, \tag{3.9}$$

and

$$SZS^t = Z. \tag{3.10}$$

Equation (3.10) defines what is known as the *symplectic* transformation [see, for instance, Chevalley (*32*); Weyl (*33*)].

3.5 QUATERNION STRUCTURE

There is a *quaternion* structure associated with the symplectic group of transformations. To express this, one introduces a unit matrix I and three 2×2 complex matrices $\tau = -i\sigma$, where the three matrices σ are the familiar Pauli spin matrices,

$$\sigma_1 = \begin{bmatrix} 0 & 1 \\ 1 & 0 \end{bmatrix}, \quad \sigma_2 = \begin{bmatrix} 0 & -i \\ i & 0 \end{bmatrix}, \quad \sigma_3 = \begin{bmatrix} 1 & 0 \\ 0 & -1 \end{bmatrix}. \tag{3.11}$$

The matrix Z can then be written as $Z = \tau_2 I$,

$$Z = \begin{bmatrix} 0 & -1 & & & & & \\ 1 & 0 & & & & & \\ & & 0 & -1 & & & \\ & & 1 & 0 & & & \\ & & & & 0 & -1 & \\ & & & & 1 & 0 & \\ & & & & & & \ddots \end{bmatrix}, \tag{3.12}$$

consisting of 2×2 blocks

$$\begin{bmatrix} 0 & -1 \\ 1 & 0 \end{bmatrix}, \tag{3.13}$$

along the leading diagonal, all other elements of Z being zeros.

A detailed analysis of the present case (see Chapter 8) shows that the property of being "quaternion real" is characteristic of the Hamiltonian of a system with odd half-integral total angular momentum subject to a nonisotropic external field (like an external electric field) which does not destroy the time-reversal invariance of the Hamiltonian.

3.6 THE THREEFOLD WAY OF INVARIANCE

One can thus summarize the situation by saying [Dyson (*34*)] that there is a *three-fold way* of utilizing the orthogonal, the unitary, and the symplectic groups as the canonical transformation group that is compatible with the invariance of the Hamiltonian resulting from isotropy and homogeneity of space-time [see Table III-2, Porter (*19*)].

In the next chapter, we introduce and discuss the Gaussian and the orthogonal ensembles.

Table III - 2

Time-reversal symmetry	Rotational symmetry	Hamiltonian	Canonical group
Good	Good	Real	Orthogonal
Good	Not good, but spin integral	Real	Orthogonal
Good	Not good, but spin half-integral	Quaternion real	Symplectic
Not good	Good, or not good	Complex	Unitary

CHAPTER 4
GAUSSIAN AND ORTHOGONOL ENSEMBLES

4.1 THE GAUSSIAN ENSEMBLE

The *Gaussian* ensemble, denoted by E_G, was introduced by Wigner [*35-38*] and is characterized by a Hamiltonian which is a real symmetric matrix H_{ij}, where $i, j = 1, \ldots, N$, and N is a fixed integer. One thus has $N(N + 1)/2$ matrix elements which are assumed to be independent Gaussian random variables with the joint distribution function

$$D(H_{ij}) = A \exp\left[-\left(\sum_i H_{ii}^2 + 2\sum_{i<j} H_{ij}^2\right)\Big/ 4a^2\right], \tag{4.1}$$

where A and a are constants. Thus one assumes that each system with N quantum states occurs in the ensemble E_G with a statistical weight given by Eq. (4.1).

The reality of the Hamiltonian ensures that the system is time-reversal invariant. It can be shown [Porter and Rosenzweig (*39*)] that Eq. (4.1) is a consequence of two assumptions:

(1) The components H_{ij} are statistically independent; and

(2) The function $D(H_{ij})$ is invariant under the transformation $H \to R^{-1}HR$, where R is a *real orthogonal* matrix. Although the assumption (2) is natural in order to give equal weight to all kinds of interactions, the definition of the Gaussian ensemble E_G is nevertheless somewhat arbitrary since assumption (1) is artificial and without physical motivation.

The unsatisfactory feature of the distribution function (4.1) is that one cannot define a uniform probability distribution on an infinite range, and hence a restriction on the magnitudes of H_{ij} has to be made since otherwise it is impossible to define an ensemble in terms of H_{ij} in which all interactions are equally probable.

4.2 THE ORTHOGONAL ENSEMBLE

The *orthogonal* ensemble [Dyson (*4*)], denoted by E_1, is defined with a slight change from the Gaussian ensemble E_G. A system is represented in E_1 not by its Hamiltonian H but by an $N \times N$ unitary matrix S whose eigenvalues are N complex numbers $\exp(i\theta_j)$, where $j = 1, \ldots, N$, distributed around the unit circle. The matrix S is a function of the Hamiltonian H so that the angles θ_j are functions of the energy levels E_j of the system, and over a small range of angles, θ_j is linear in E_j.

The basic statistical hypothesis here is that the behavior of n consecutive levels of an actual system, where n is small compared with the total number of levels, is statistically equivalent to the behavior in the ensemble E_1 of n consecutive angles θ_i on the unit circle, where n is small compared with N.

Both the Gaussian ensemble and the orthogonal ensemble are restricted to $N \times N$ matrices, and are mutilations of an actual nucleus with an infinite number of energy levels. The most one can ask of such an ensemble is that it correctly reproduce level distributions over an energy range small compared with the total energy of excitation. The connection between the matrix S and the Hamiltonian is vague. The energy level distributions predicted by the Gaussian ensemble and by the orthogonal ensemble are unrealistic.

4.3 THE SEMICIRCLE DISTRIBUTION

The first gives the so-called *semicircle distribution* [Wigner (*37*)]

$$P(E) = (2\pi Na^2)^{-1} (4Na^2 - E^2)^{1/2}, \quad E^2 < 4Na^2, \tag{4.2a}$$

and

$$P(E) = 0, \quad E^2 > 4Na^2, \tag{4.2b}$$

which does not resemble the level distribution of a nucleus. The distribution (4.2) is also very different from that of the real roots of an algebraic equation of order N.

Figure 4.1 is a histogram of $P(E)$ obtained by diagonalizing 20×20 matrices, selected at random from a certain ensemble. As can be seen, the distribution approaches a semiellipse. In fact a semicircle is actually a misnomer since the two axes do not even have the same dimensions [Wigner (*6*)].

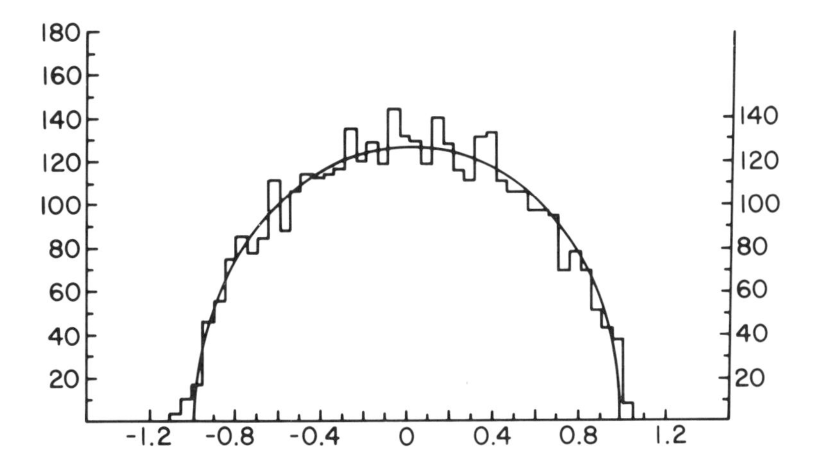

Figure 4.1 The histogram of the semicircle distribution obtained by diagonalizing 197 real symmetric 20×20 matrices [Wigner (*6*)].

As has been pointed out before, the distribution (4.2) shows no similarity to the observed distribution in spectra. The behavior at large positive E is not relevant since what is known, and what could be hoped to be reproduced by the ensemble, is the distribution in the neighborhood of the lowest energy level.

The density in the neighborhood of the lowest state in nuclei, for example, is such that there are few levels per million electron volts. Around 5 meV, on the other hand, there are several levels in an interval of 100 eV. There seems to be an exponential increase with energy. The density of the levels as a function of the energy is convex from below, whereas the semicircle or semiellipse is concave. It should be surmised that the convex distribution applies only in the neighborhood of the lower range of the asymptotic formula in the region where the asymptotic formula does not hold. The density in the range of the semicircle law is proportional to $N^{1/2}$, where N is the dimension of the random matrix. If it were proportional to a lower power of N outside the ellipse, this would not show in the asymptotic law but might explain the region in which the density of levels increases fast.

4.4 THE WISHART ENSEMBLE

This region was discussed in more detail for the so-called *Wishart ensemble* [Baronk (*40*)], an ensemble in which the matrix elements are independent of each other and each shows a Gaussian distribution. It was found that the semicircle law is too accurate. On the average there are only about two levels outside its range. It is clear that the existence of a reasonably large region in which the second energy derivative of the density of levels is positive does not follow from the assumptions made so far.

The orthogonal ensemble gives in the large a uniform distribution around the unit circle. Although the orthogonal distribution, like the Gaussian distribution, is unphysical, it has the advantage of simplicity and absence of spurious end effects.

The precise definition of the orthogonal ensemble E_1 is as follows: A system is characterized by an $N \times N$ symmetric unitary matrix S. Since the space T_1 of all matrices S is compact, it makes sense to require that the ensemble E_1 contain all possible S with equal probability.

To give a meaning to equal probability, one requires a measure μ in the space T_1. Since the matrices S do not form a group, the definition of μ is not easy, and is done as follows: One writes S as $U^t U$, where U is a unitary matrix. An infinitesimal neighborhood of S in T_1 is given by

$$S + dS = U^t (1 + i\, dM)\, U, \tag{4.3}$$

where dM is a real, symmetric, infinitesimal matrix with elements dM_{ij}, and dM_{ij}, with $i \leqslant j$, vary independently through some small intervals of lengths $d\mu_{ij}$. The measure of this neighborhood is then defined as

$$\mu(dS) = \prod_{i<j} d\mu_{ij}. \tag{4.4}$$

The ensemble E_1 is then defined by demanding that the probability that a system of E_1 belongs to the volume element dS is

$$P(dS) = (V_1)^{-1}\, \mu(dS), \tag{4.5}$$

where V_1 is the total volume of the space T_1,

$$V_1 = \int \mu(dS). \tag{4.6}$$

It can be shown that $\mu(dS)$ is independent of the particular U that was chosen to represent S.

4.5 INVARIANCE AND EQUAL PROBABILITY

It also follows that for fixed S the unitary matrix U is undetermined precisely to the extent of a transformation

$$U \to RU, \tag{4.7}$$

where R is an arbitrary real orthogonal matrix. In addition, one can show that the ensemble E_1 is uniquely defined in the space T_1 of symmetric unitary matrices by the property of being invariant under every automorphism $S \to W^t S W$ of T_1 into itself, where W is any unitary matrix. Thus one states in mathematical language the precise meaning of the vague statement "all systems occur in E_1 with equal probability."

The automorphism $S \to W^t S W$ is not a mere change in the representation of states. It is a physical alteration of the system S into a different system. One may visualize S as representing an unknown system enclosed in a "black box," and S is the transformation matrix of the system from some initial state ψ_i to a final state ψ_f.

The transformations $S \to W^t S W$ then means that one subjects the initial state to some further interaction W, and the final state to the same interaction W^t in a time-symmetric way. If we are totally ignorant of the interactions occurring inside the black box, the additional interaction W cannot increase or decrease our ignorance, and if all systems S were equally probable at the beginning, the application of W must not change that. Invariance of the ensemble E_1 under the transformation $S \to W^t S W$ is a mathematical idealization of the hypothetical state of total ignorance.

It remains only to justify, on physical grounds, the choice of the basic space T_1 of symmetric unitary matrices. Alternative choices are discussed in Chapter 8. The choice of T_1 has the same motivation as the choice of real symmetric matrices for the Gaussian ensemble E_G. Symmetric unitary matrices are physically appropriate under two alternative conditions:

(1) The systems are invariant under time inversion and space rotations; or

(2) The systems are invariant under time inversion and contain an even number of half-integer spin particles.

The symmetry of the S matrix for systems satisfying condition (1) has been proven in a particularly simple case [Coeseter (*41*)]. In the neutron capture resonances case, for example, condition (1) always holds and the ensemble E_1 is the one to be used.

In the next chapter, the unitary ensemble will be discussed.

CHAPTER 5
UNITARY ENSEMBLE

5.1 SYSTEMS WITHOUT TIME-REVERSAL SYMMETRY

In Chapter 4 we discussed the Gaussian and orthogonal ensembles, and in Chapter 8 we will discuss the symplectic ensembles. We now briefly discuss systems without time-reversal symmetry, a unitary ensemble [Dyson (*4*)].

The unitary ensemble, denoted by E_2, is a simple one. From the physical point of view such a system is easily created, for example, by putting an ordinary atom or nucleus into an externally generated magnetic field, provided the splitting of the levels by the magnetic field is of the same order of magnitude as the average level spacing in the absence of the field. The magnetic interaction must in fact be so strong that it completely mixes up the level structure which would exist in the zero field. This situation does not occur in nuclear physics, but it could occur in atomic or molecular physics.

The Hamiltonian of a system without invariance under time reversal is usually an arbitrary Hermitian matrix which is not restricted to the symmetric or self-dual. The system is represented by an $N \times N$ unitary matrix S belonging to the space T_2 of all unitary matrices. Since the space T_2 is the unitary group $U(N)$, it is a simple matter to define a uniform ensemble E_2 in T_2, and an invariant group measure in $U(N)$ can easily be established.

The ensemble E_2 is then defined as follows: A neighborhood of S in T_2 is given by

$$S + dS = U(1 + i\,dH)\,V, \tag{5.1}$$

where U and V are two unitary matrices such that $S = UV$, dH is an infinitesimal Hermitian matrix with elements $dH_{ij} = dH_{ij}^1 + i\,dH_{ij}^2$, and dH_{ij}^1 and dH_{ij}^2 vary independently through small intervals of lengths $d\mu_{ij}^1$ and $d\mu_{ij}^2$, respectively.

The invariant group measure $\mu(dS)$ is defined by

$$\mu(dS) = \prod_{i,j} d\mu_{ij}^1\, d\mu_{ij}^2, \tag{5.2}$$

independently of the choice of the two matrices U and V. The ensembles E_2 give each neighborhood dS the statistical weight

$$P(dS) = (V_2)^{-1}\, \mu(dS), \tag{5.3}$$

where V_2 is the volume of the space T_2. The unitary ensemble E_2 is uniquely defined in the space T_2 of unitary matrices by the property of being invariant under every automorphism $S \to USW$ of T_2 into itself, where U and W are two matrices of the space T_2.

In the rest of this chapter we will discuss, following Brody, Flores, French, Mello, Pandey, and Wong (*42*), the problem of the level repulsion in spectra. More accurate mathematical formulation of this topic will be given in the next chapter.

5.2 LEVEL REPULSION IN SPECTRA

For a random sequence, the probability that an energy level will be in the infinitesimal interval

$$(E+S,\ E+S+dS), \tag{5.4}$$

proportional, of course, to dS, is independent of whether or not there is a level at E. This result is changed if one introduces the concept of level repulsion.

Given a level at E, let us denote by $P(S)dS$ the probability that the next level ($S \geqslant 0$) be in the interval (5.4). We then have for the nearest-neighbor spacing distribution the following formula [Brody *et al.* (*42*)]:

$$P(S)dS = P(1 \in dS \,|\, 0 \in S)\, P(0 \in S). \tag{5.5}$$

Here $P(n\epsilon S)$ is the probability that the interval of length S contains n levels, and $P(n \in dS \,|\, m\epsilon S)$ is the conditional probability that the infinitesimal interval of length dS contains n levels when that of length S contains m levels.

The first term on the right-hand side of Eq. (5.5) is dS times a function of S which we denote by $r_{10}(S)$, depending explicitly on the choices 1 and 0 of the discrete variables n and m. The second term on the right-hand side of Eq. (5.5) is given by

$$\int_S^\infty P(x)\, dx, \tag{5.6}$$

the probability that the spacing is larger than S. Accordingly, we obtain

$$P(S) = r_{10}(S) \int_S^\infty P(x)\, dx, \tag{5.7}$$

whose solution can easily be found to be given by

$$P(S) = Cr_{10}(S)\, e^{-\int^S r_{10}(x)dx}, \tag{5.8}$$

where C is a constant.

The Poisson law then follows if one takes

$$r_{10}(S) = \frac{1}{D}, \tag{5.9}$$

where D is the mean local spacing, so that $1/D$ is the density of levels. Wigner's law then follows by assuming a *linear* repulsion,

$$r_{10}(S) = \alpha S, \tag{5.10}$$

where α is a constant. The two constants C and α can then be determined from the conditions

$$\int P(x)\,dx = 1, \tag{5.11a}$$

and

$$\int xP(x)\,dx = D. \tag{5.11b}$$

One then finds that

$$P(S) = \frac{1}{D}\,\mathrm{e}^{-S/D}; \quad S \geqslant 0, \tag{5.12}$$

for the *Poisson distribution* and

$$P(S) = \frac{\pi S}{2D^2}\,\mathrm{e}^{-\pi S^2/4D^2}; \quad S \geqslant 0, \tag{5.13}$$

for *Wigner's distribution.*

Equation (5.13) displays the repulsion explicitly since $P(0) = 0$. This is in contrast to the Poisson form (5.12) which has a maximum at $S = 0$. The Wigner distribution (5.13), like that of Poisson, is a standard one in statistics. It is the distribution for the square root of the sum of the squares of two independent Gaussian random variables of type $G(0, D\sqrt{2/\pi})$, where $G(a, b)$ has centroid a and variance b^2, and is sometimes called the *Rayleigh distribution.* Hence, one has

$$\rho(x) = \frac{1}{\sqrt{2\pi b^2}}\,\mathrm{e}^{-(x-a)^2/2b^2} \tag{5.14}$$

for the corresponding probability density.

The assumption that $r_{10}(S) = \alpha S$, which was used to derive the Wigner distribution (5.13), leads to great difficulties (compare Chapter 6). In the first place, this unbounded linear form for $r_{10}(S)$ must be inappropriate for large S. But even for small S why should one assume a repulsion?

Nevertheless, one can show that there are some simple plausibility arguments for this linear form of the function $r_{10}(S)$, even though the result cannot be correct for every system. This is so since one can almost always construct a Hamiltonian which has a given spectrum and hence a very different spacing law from either of the Poisson or the Wigner distributions. But evidently, certain Hamiltonians are in some sense more "likely" to occur than others. In Wigner's distribution, there are thus tacit assumptions about the relative probability of different Hamiltonians.

When one introduces an ensemble of Hamiltonians, a fundamental difficulty seems to arise since there appears to be no natural and significant weighting function for such an ensemble. In particular, there seems to exist no equivalent to the Liouville theorem which provides such a weighting function in the theory of classical statistical mechanics.

Many different forms for such a function are possible, depending on the quantum numbers which one takes to be exactly conserved and on the importance which one attaches to various features of the Hamiltonian, such as its two-body nature, etc.

It follows that a probability argument which pays no attention to the specific features of the Hamiltonian cannot explain the nature, the origin, or the consequences of the level repulsion. It should also be noted that the above discussion on the level repulsion applies only for a *pure sequence*, namely, one whose all levels have the same values of the exact quantum numbers. Cases of *mixed sequences* will be discussed in the sequel.

Figure 5.1 shows a set of spectra, where runs of 50 levels taken from three very different sources, were brought together. These are the slow-neutron resonance region of ^{167}Er, the neighborhood on an isobaric-analogue state in ^{49}V, and a section of a large shell-model calculation with a realistic interaction [Brody *et al.* (*42*)].

The levels in each spectrum have the same (J, P), and the scales have been chosen so that the average spacing D is the same for each. The spectra in these energy levels are similar in their general nature.

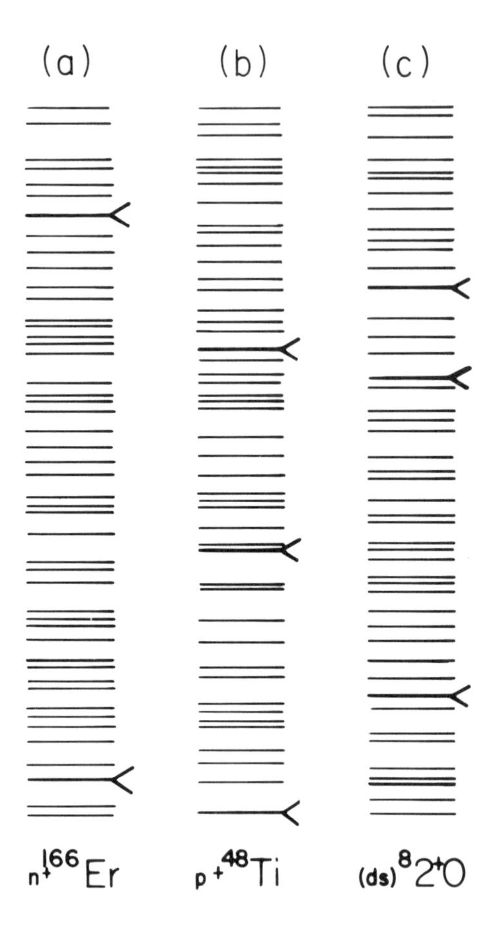

Figure 5.1 Segments of complex spectra, each contains 50 levels and rescaled to the same spectrum span [Brody *et al.* (*42*)]. The first two spectra show experimental results for neutron and proton resonances, whereas (c) shows the central region of a 1206-dimensional, $J = 2^+$, $T = 0$, shell-model spectrum. In all three cases, all the states have the same exact symmetries. The "arrowheads" denote the occurrence of pairs of levels with spacings smaller than one quarter of the average.

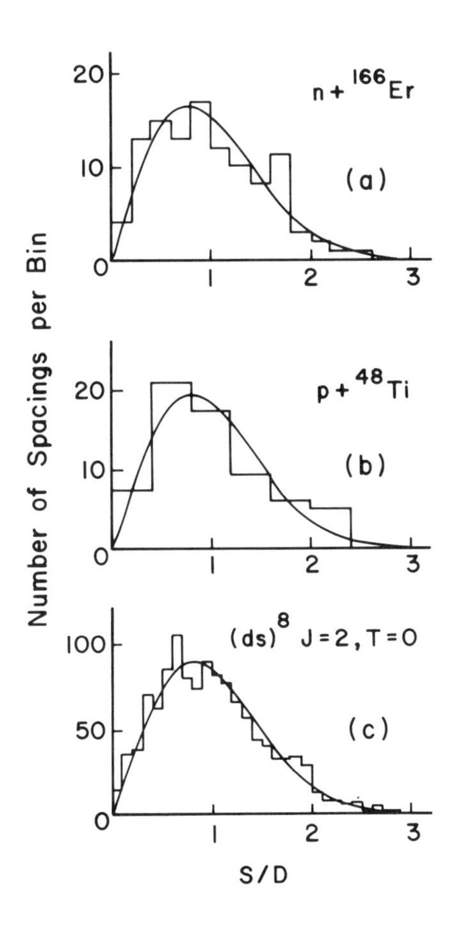

Figure 5.2 Nearest-neighbor spacing histograms for the three cases of Figure 5.1, constructed by considering all the available levels instead of the 50 used in Figure 5.1 [Brody *et al.* (*42*)]. The spacings S are expressed in terms of the local spacing unit D. The Wigner distribution (5.13) is shown for the three cases.

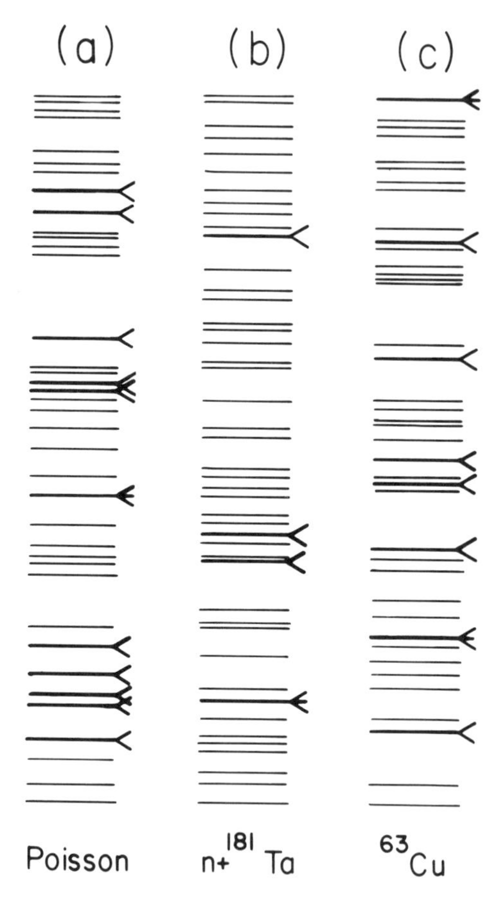

Figure 5.3 Segments of complex spectra, each contains 50 levels and rescaled to the same spectrum span [Brody *et al.* (*42*)]. The first spectrum shows a Poisson sequence, while the spectra (b) and (c) are spectra with mixed exact symmetries, the first an experimental spectrum with $J = 3^+, 4^+$ and the second a shell model spectrum with $J = \frac{1}{2}^-, \frac{3}{2}^-, \ldots, \frac{19}{2}^-$. The "arrowheads" mark the occurrence of pairs of levels with spacings smaller than one quarter of the average.

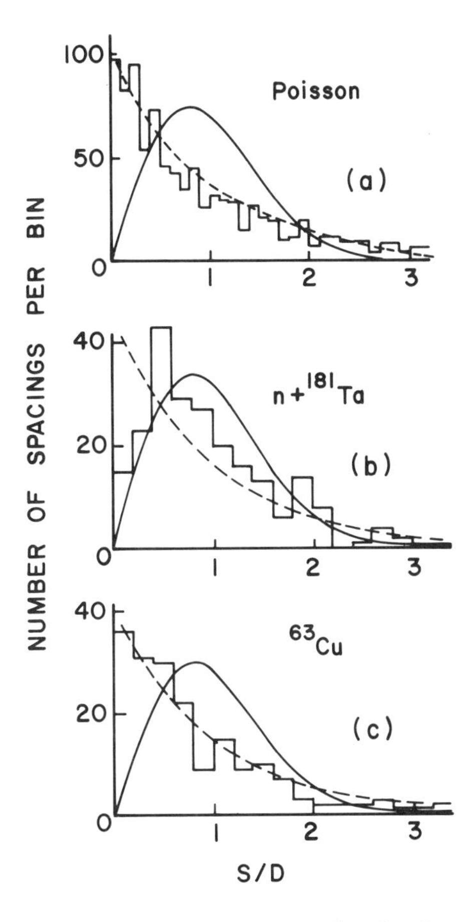

Figure 5.4 Nearest-neighbor spacing histograms for the three cases of Figure 5.3, constructed by considering all the available levels and not only the 50 levels used in that figure [Brody *et al.* (*42*)]. Spacings S are expressed in terms of the local spacing unit D. The Wigner and the Poisson distributions are shown for all three cases.

For example, the number of spacings which are much smaller than the average spacing is statistically the same for all the three cases. (In the figures, this is made clear by marking those levels which are smaller than $D/4$ by arrowheads.)

The similarity of the spectra is made even more clear in Figure 5.2, which gives the histograms of the nearest-neighbor spacings for each of the spectrums in Figure 5.1. As can be seen, they resemble each other closely enough that one may regard the three spectra of Figure 5.1 as having the same nearest-neighbor spacing distributions, thus exhibiting a regularity of statistical nature, even though the spectra themselves differ in the three examples. The continuous curves shown in the figure represent the Wigner distribution discussed above. As is seen, the Wigner distribution fits the three cases quite well.

The small spacings of our examples have a small probability of occurrence. So is the case for large spacings. The limiting case in which there are neither small nor large spacings is the *uniform spectrum* which might be described as *rigid*. Most spectra of energy levels usually have a high degree of rigidity.

Assuming that the individual states have complicated structures, one can further postulate that the levels should form a completely random sequence which can be visualized as being distributed in energy as the pulses from a radioactive target are distributed in time. In that case, as was shown by Eq. (5.12) above, the spacing distribution would have been of the Poisson type, in which small spacings predominate. This is shown in the random sequence spectrum of Figure 5.3 (case a) and its corresponding histogram given in Figure 5.4 (case a). As is easily noted, the latter example is very different from the examples given in Figures 5.1 and 5.2 which, by comparison with the random case, may be regarded as displaying a level repulsion. (See the next chapter for a rigorous definition of the concept of repulsion.)

As was mentioned before, the above discussion on the level repulsion applies only for a pure sequence, i.e., a sequence whose levels all have the same values of the exact quantum numbers. Figure 5.3 (cases b and c) shows mixed sequences. The figure shows a segment of the ^{182}Ta spectrum, derived from the slow neutrons on ^{181}Ta for which $J = 7^+/2$, and a low-lying region deriving from a ^{63}Cu shell-model calculation. In the first spectrum, there are interwoven runs of levels with $J = (3^+, 4^+)$ and, in the second, with $J = (1^-/2 - 19^-/2)$.

In these cases, the level repulsion is moderated by the vanishing of the Hamiltonian matrix elements connecting two different J values, the spectra have moved toward random, and the spacing distributions of Figure 5.4 (cases b and c) toward the Poisson distribution.

If one considers the different J sequences as essentially independent, and each one is described by the Wigner distribution (5.13), it is then possible (see next Chapter) to calculate the resulting distribution. One can, therefore, assume that the fundamental problem is that of understanding the pure sequence, since all the others may be obtained from it by superposition.

Yet an unsolved problem is to understand why different sequences should be independent. After all, they are derived from the same Hamiltonian.

5.3 REPULSION IN LOW ENERGY SPECTRA

The spectra discussed above represent high excitation energy levels. In order to relate them to the low energy spectra, one spans the gap, as was described above, by making use of large shell-model calculations.

Figure 5.5 (a) shows the spacing histogram for levels 10-110 of the 1206-dimensional shell-model spectrum from which levels 576-625 were used in Figure 5.1, case c. It displays a level repulsion similar to that shown in Figure 5.2, case c, for the entire spectrum, showing that the same spacing distribution applies over a wide range of energies.

The same is true, in fact, for other fluctation properties—though certain small differences will appear on detailed examination. The results were verified more explicitly by comparing separate segments of the spectrum.

However, one cannot conclude from this that the fluctuation pattern extends all the way to the ground state, because of the rapid secular variation of the density which occurs at the extremes of the

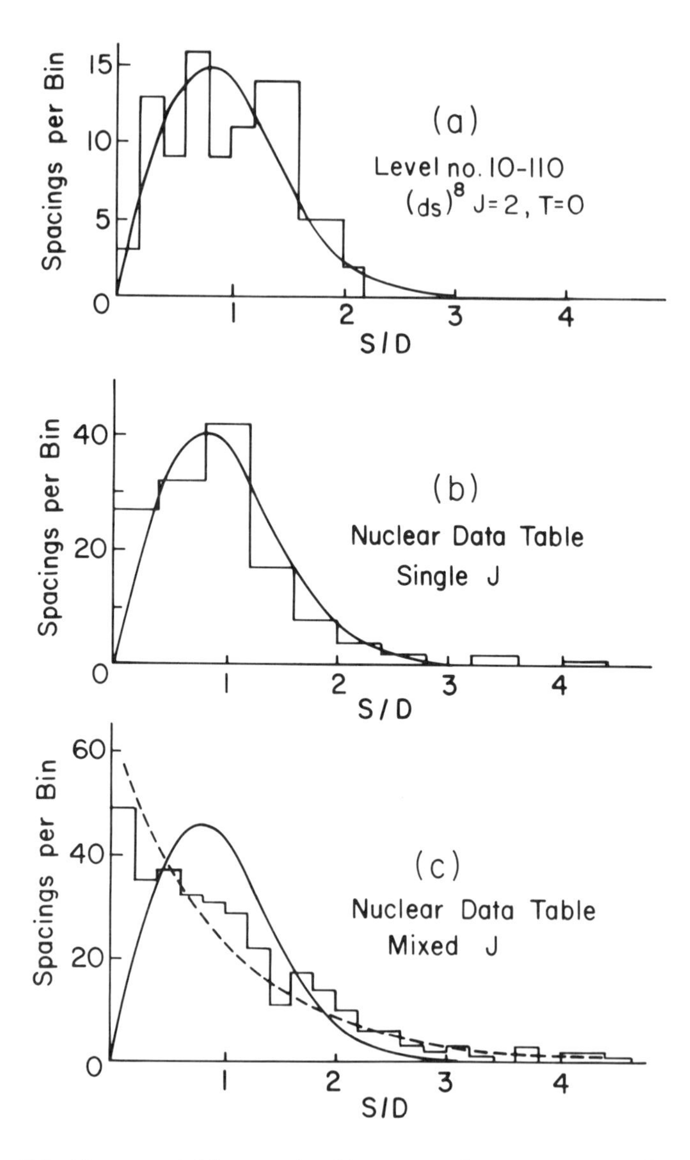

Figure 5.5 Nearest-neighbor spacing historgrams [Brody *et al.* (*42*)]: (a) for a low-lying segment of the 1206-dimensional shell-model spectrum used in Figures 5.1 and 5.2, cases c; (b) for the nuclear data table, considering spacings between the ground state and the lowest excited state of the same exact symmetries; (c) for the nuclear table considering the lowest spacing irrespective of the exact symmetries.

spectrum. (Shell-model examples of this sort will be given in the sequel.) But one can consider the ground-state fluctuations in a different manner by plotting the spacings between the two lowest levels of the same J and P for different nuclei as a function of the mass number A [Brody *et al.* (*42*)].

It then follows that except for the very light nuclei, a spacing, locally averages in A, can now be determined quite consistently. In fact, it is found to be given by $1/A$.

Nuclei can be divided into two types: The first group consists of the even-even, rotational, magic, and doubly magic nuclei (which might be roughly described as "collective"), for which the spacings differ only marginally from the average; and the second group which includes all the other nuclei and is called the main group of nuclei. After normalizing the data so that the average spacing is constant, the second group of nuclei yields a spacing histogram roughly of Wigner's type, displaying the level repulsion seen in the slow-neutron resonance region. This property is shown in Figure 5.5, case b, (which should be compared, for instance, with Figure 5.2, case a).

On the other hand, if the normalized spacings between ground and first-excited states are taken irrespective of the good quantum numbers involved, the spacing distribution is then comparable with the three cases given in Figure 5.4. This is shown in Figure 5.5, case c, and is of the Poisson's type.

The above results involve a statistical property which covers the entire range of energies of the discrete nuclear spectra. They were investigated not by examining many spacings taken from a single spectrum, but rather by combining spacings, each one taken from the series of different nuclei.

What was done here is analogous to the standard procedure in the theory of classical statistical mechanics. Whenever the time average of a function for an individual system is not calculable, one constructs a large set of theoretical replicas of the system and takes the average over this set.

As was discussed in details in Chapter 1, for complex systems, the introduction of ensembles is essential since it is not reasonable to integrate the equations of motion, namely, to construct and diagonalize a sufficiently extended shell-model matrix, even though computer calculations are possible and important in supplementing theoretical investigations. However, computer calculations are not substitute if only because one can never be certain that the results from relatively small matrices are properly "asymptotic" or even applicable to very complex nuclei.

It is also worthwhile mentioning that the equivalence of phase and time averaging in statistical mechanics defines the essence of the "ergodic" and related problems. These have their counterpart in the statistical theory of energy levels (see Appendix C). Nevertheless, it is worthwhile remembering that the nearest-neighbor spacings, at least, have the same distribution for the main group of nuclei (see above) when taken along an individual spectrum a million levels above the ground state as when taken across an ensemble at the ground state itself. This is clearly an evidence for some combination of ergodicity and stationarity.

On the other hand, the spectra of the excluded nuclei are dominated by collectivities or symmetries (or both). Since these are not effective at high excitation, the equivalence of the spacing laws found above for the main group is not to be expected for these nuclei. In more general treatments of fluctuations, one should be able to take these nuclei into account also.

Finally, it is worthwhile emphasizing that the statistical theory of energy levels is formulated along the energy axis rather than along the time axis as in the conventional theory of statistical mechanics (see also Chapter 1). The analog of time is therefore energy. This analogy, however, is not perfect because while thermodynamical systems evolve along the time axis, here one has no way of describing mathematically the transition from one level to the next.

Moreover, since it is mainly the region of discrete energy levels or narrow resonances which is of interest, one can limit himself to the matrix representations of the Hamiltonians. It then follows that only a small low-lying part of the spectrum can have a direct one-to-one correspondence (as opposed to a statistical one) between the model states and those of the physical system.

In the next chapter, we will discuss the eigenvalue-eigenvector distributions of the Gaussian ensemble.

CHAPTER 6
EIGENVALUE-EIGENVECTOR DISTRIBUTIONS OF THE GAUSSIAN ENSEMBLE

6.1 THE GAUSSIAN ENSEMBLE

The detailed calculation of the measure and the matrix element distribution for the Hamiltonian submatrix H corresponding to one set of symmetry labels (like the spin J and the parity P) discussed in the previous chapters leads to (see Chapter 9)

$$d\mu(H_\beta) = 2^{\beta N(N-1)/4} \left[\prod_{\mu=1}^{N} dH_{\mu\mu} \right] \left[\prod_{\mu<\nu=1}^{N} \prod_{k=0}^{\beta-1} dH_{k\mu\nu} \right], \tag{6.1}$$

or

$$d\mu(H_\beta) \cong \left[\prod_{\lambda<\mu=1} |E_\lambda - E_\mu|^\beta \right] d\mu(H_D) \times \begin{cases} d\mu(R) \\ d\mu(U) \\ d\mu(S) \end{cases} \tag{6.2}$$

for the measure, and to

$$P(H, \beta) = \frac{e^{-[\mathrm{Tr}(H-E_0 I)^2]/4a^2}}{(4\pi a^2)^{[N+\beta N(N-1)/2]/2}}, \tag{6.3}$$

along with

$$\int P(H;\beta)\, d\mu(H_\beta) = 1, \tag{6.4}$$

for the matrix element distribution, where $\beta = 1$ corresponds to the orthogonal, $\beta = 2$ to the unitary, and $\beta = 4$ to the symplectic cases.

These formulas, however, are not directly useful to obtaining numerical results for comparing with energy level spacings, widths, and expectation values. From these equations, however, one can obtain the eigenvalue distributions $P(E_1, \ldots, E_N;\beta)$ and the eigenvector distributions $P(R)$, $P(U)$, or $P(S)$ according to the symmetry of the problem. Since the distribution function $P(H)$ is invariant and the Jacobian [Chapter 9, Eq. (9.12)] factors in the eigenvalues and the measures, it follows that the eigenvectors are statistically *independent* of the eigenvalues in the Gaussian ensembles.

Hence there is no statistical correlation between the eigenvalue spacings and the widths or expectation values. [Some data suggest a small correlation, however. See Grag, Rainwater, Petersen, and Havens (*43*).] One has for the eigenvalue distributions

$$P(E;\beta) \cong \exp\left[-\frac{1}{4a^2} \sum_{\lambda=1}^{N} (E_\lambda - E_0)^2 \right] \left[\prod_{\lambda<\mu=1}^{N} |E_\lambda - E_\mu|^\beta \right], \tag{6.5}$$

whereas the eigenvector distributions $P(R)$, $P(U)$, and $P(S)$ are constants proportional to the inverse of the total integrals of the measures $d\mu(R)$, $d\mu(U)$, and $d\mu(S)$, respectively.

Hence the eigenvectors form a set of random, mutually perpendicular vectors, and therefore one can obtain the single-eigenvalue distribution $P(E;\beta)$ corresponding to the level density in some sense, the spacing distributions $P_N{}^k(\chi;\beta)$, where $\chi = S/D$ and S is the spacing while D is the mean spacing between levels. A complete unified treatment for the spacing distributions and the single-eigenvalue distribution is not available.

Example. The Gaussian ensembles of 2 X 2 matrices. From Eq. (6.5) one obtains

$$P(E_1, E_2; \beta) = C_\beta |E_1 - E_2|^\beta \times e^{-[(E_1-E_0)^2 + (E_2-E_0)^2]/4a^2}, \tag{6.6}$$

where C_β is a normalization factor given by

$$C_\beta = 1/(2a)^{2+\beta} 2^{\beta/2}\, \Gamma[(1+\beta)/2]\ \Gamma(1/2). \tag{6.7}$$

The mean spacing is given by

$$D = a2^{3/2} \frac{\Gamma[1 + (\beta/2)]}{\Gamma[(1+\beta)/2]}, \tag{6.8}$$

and the nearest-neighbor spacing distributions are given by

$$P_2{}^0(\chi; 1) = \frac{\pi}{2} \chi e^{-\pi\chi^2/4} \quad ; \text{(orthogonal)}, \tag{6.9}$$

$$P_2{}^0(\chi; 2) = \frac{32}{\pi^2} \chi^2 e^{-4\chi^2/\pi} \quad ; \text{(unitary)}, \tag{6.10}$$

$$P_2{}^0(\chi; 4) = \frac{2^{18}}{3^6\pi^3} \chi^4 e^{-64\chi^2/9\pi} \quad ; \text{(symplectic)}. \tag{6.11}$$

Here $\chi = S/D$, where S is the spacing and D is given by Eq. (6.8).

6.2 THE REPULSION OF LEVELS

In general, the characteristic values of a real symmetric matrix, or a complex Hermitian matrix, "repel" each other [von Neumann and Wigner (*44*)]. This means that if the matrix elements depend on a number of continuous parameters, then the dimension of the domain in the space of the parameters for which the matrix has a double root is, in general, lower by two than the dimension of the parameter space itself.

To see this, we go back to the case of 2×2 matrices discussed above. Let the symmetric matrix be given by

$$\begin{bmatrix} a & b \\ b & c \end{bmatrix}. \tag{6.12}$$

Then the roots (characteristic values) of this matrix are given by the solution of the quadratic equation

$$E^2 - (a + c)E + (ac - b^2) = 0. \tag{6.13}$$

Hence they are

$$E = \frac{1}{2}(a + c) \pm \frac{1}{2}[(a - c)^2 + 4b^2]^{1/2}. \tag{6.14}$$

As can easily be seen, the two roots are equal if $a = c$ and $b = 0$. Now the space of parameters of the matrix (6.12) is three dimensional. In this space, whose coordinates are given by a, b, and c, the equations $a = c$ and $b = 0$ represent a straight line in the $b = 0$ plane, which has one dimension, of course. It follows from this property of matrices that the probability for a spacing S (interval between adjacent roots) is proportional to S itself if S is very small compared with the average spacing D.

The property of a "repulsion" of energy levels (characteristic values, or roots of the matrix) can also be seen from Eqs. (6.9)-(6.11), which shows how the three ensembles are reflected in the nearest-neighbor spacing distribution. For small $\chi = S/D$ the exponential parts of Eqs. (6.9)-(6.11) approach one and hence the distributions functions vary as χ^β for the three cases. This is in contrast to an ordered sequence of random numbers following the Poisson distributions $P^0(\chi) = \exp(-\chi)$ [see Eq. (2.3)], which approaches a constant as χ goes to zero. Hence, there is a repulsion of energy levels associated with an absence of small spacings (namely absence of degeneracy) [Landau and Smorodinsky (*45*)].

The behavior of Eqs. (6.9)-(6.11) for large spacings is also interesting in the two-dimensional case. The result, when $\chi \to \infty$, is given by

$$\ln P_2{}^0(\chi;\beta) \to \text{const} - \left[\frac{\Gamma(1+\beta/2)}{\Gamma[(1+\beta)/2]}\right]^2 \chi^2 + \beta \ln \chi. \qquad (6.15)$$

In the case of large dimension where $N \to \infty$, Eq. (6.15) has to be modified by the addition of a linear term [Dyson (*46*)]. But Eqs. (6.9)-(6.11) do not differ numerically very much from the infinite-dimensional results [Gaudin (*47*)].

6.3 WIGNER'S CONJECTURE

It is interesting to note that Eq. (6.9) is identical with a conjecture made by Wigner in 1957 for the spacing distribution [Wigner (*48*)]. The formula was supposed to apply to a series of levels having the same values of all identifiable quantum numbers, such as angular momentum and parity. It is very well supported by experimental data and by numerical tests with random matrices of high order. However, it is known to be false.

The correct distribution function was obtained later by Mehta and Gaudin (*49, 50*), who computed it numerically. They found that Wigner's distribution function is not identical with the one they derived but is surprisingly close to it, the difference being less than 0.0162 over the whole range of $\chi = S/D$. Hence for practical purposes Wigner's formula is justified.

Within experimental uncertainties, changing the dimension N in the Gaussian ensemble does not affect significantly the form of the predicted nearest-neighbor spacing distribution. In fact almost every other approach to the nearest-neighbor spacing distribution yielded results indistinguishable experimentally from Eq. (6.9).

The *distribution of the λth eigenvalue* is defined by (for the two-dimensional case)

$$P_2(E;\beta,\lambda) = 2!\,C_{2\beta}\int_{-\infty}^{\infty} dE_1 \int_{E_1}^{\infty} dE_2\, |E_1 - E_2|^\beta \times e^{-(E_1{}^2+E_2{}^2)/4a^2}\,\delta(E - E_\lambda). \tag{6.16}$$

Here $\lambda = 1, 2$. In N dimensions there are N distributions $P_N(E;\beta,\lambda)$. The *single-eigenvalue distribution* is defined as

$$P_N(E;\beta) = N^{-1}\sum_{\lambda=1}^{N} P_N(E;\beta,\lambda), \tag{6.17}$$

and can therefore be written in the two-dimensional case as

$$P_2(E;\beta) = C_{2\beta}\int_{-\infty}^{\infty} dE_1 \int_{-\infty}^{\infty} dE_2\, |E_1 - E_2|^\beta \times e^{-(E_1{}^2+E_2{}^2)/4a^2}\,\delta(E_1 - E). \tag{6.18}$$

The detailed calculation of this function for the three cases (orthogonal, unitary, and symplectic) gives

$$P_2(\epsilon;1) = \frac{1}{\sqrt{2\pi}}\,e^{-2\epsilon^2} + \frac{1}{\sqrt{2}}\,e^{-\epsilon^2}\,\epsilon\,\Phi(\epsilon), \tag{6.19}$$

$$P_2(\epsilon;2) = \frac{1}{\sqrt{\pi}}\,e^{-\epsilon^2}(1 + 2\epsilon^2), \tag{6.20}$$

$$P_2(\epsilon;4) = \frac{1}{12\sqrt{\pi}}\,e^{-\epsilon^2}(3 + 12\epsilon^2 + 4\epsilon^4), \tag{6.21}$$

where $\epsilon = E/2a$, and $\Phi(z)$ is an error function,

$$\Phi(z) = \frac{2}{\sqrt{\pi}}\int_0^z e^{-t^2}\,dt. \tag{6.22}$$

The *level density* is then given by $\rho_N(E;\beta) = NP_N(E;\beta)$, where $P_N(E;\beta)$ is the single-eigenvalue distribution defined above, and one has $\rho_N \to 1/D$ in the limit when $N \to \infty$, where D is given by Eq. (6.8). However, one finds in the case of two dimensions at $E = 0$ that $1/\rho_2(0;\beta)$ is equal to $(2\pi)^{1/2}a$, $\pi^{1/2}a$, and $4\pi^{1/2}a$ for $\beta = 1, 2$, and 4 respectively, whereas D is given by $(2\pi)^{1/2}a, 4(2/\pi)^{1/2}a$, and $16(2/\pi)^{1/2}a$ for the same cases.

6.4 MEAN LEVEL DENSITY

We now discuss in some detail the concepts of the mean level density and the nearest-neighbor spacing [Porter (*19*)]. For the Gaussian ensemble one can develop the asymptotic semicircle law of Wigner [see Chapter 4, and Refs. *36, 38* and *41*],

$$P_N(E;\beta) \underset{N\to\infty}{\to} \frac{G}{\pi a(\beta N)^{1/2}} \left(1 - \frac{E^2}{4a^2\beta N}\right)^{1/2}, \tag{6.23}$$

valid for $E \leqslant 2a(\beta N)^{1/2}$. Hence $\rho_N(E;\beta) = NP_N(E;\beta)$ is given by $\rho_N(0;\beta) \to_{N\to\infty} (1/\pi a)\,(N/\beta)^{1/2}$ for the case of $E = 0$.

In addition to the nearest-neighbor spacing distributions $P_N{}^0(\chi;\beta)$, the higher-order spacing distributions $P_N{}^k(\chi;\beta)$ (see above) are of interest for comparison with experimental results. The case for which N goes to infinity is again of great interest. This has been worked out by Dyson for the circular ensemble [Dyson (*51*); Kahn (*52*)].

A complete Monte Carlo calculation of quantities using the Gaussian orthogonal ensemble with a sample of 10,000 10 $\times$ 10 matrices was carried out by Porter (*53*), who obtained all spacing distributions up to $k = 8$ and showed that the rate of change of these distributions with dimension is small over the range of major probability. Dyson and Mehta (*54*) have developed efficient statistics, one of which is sensitive to long-range order properties of a sequence of levels, where the concept of selecting a set of N levels in sequence out of sequence of N levels in the circular orthogonal ensemble was developed to meet the experimental results.

6.5 REMARK

Finally some remarks should be made on the unitary and symplectic ensembles [Porter (*19*)]. In the case of the circular unitary ensemble [Dyson (*4*)] a sequence of energy levels can be obtained from the circular orthogonal ensemble by superposing two energy level sequences having the same mean spacing [Gunson (*55*)]. A sequence of energy levels in the circular symplectic ensemble can also be generated from a sequence of levels in the circular orthogonal ensemble by selecting alternate levels [Mehta and Dyson (*56*)]. It is conjectured that similar relationships hold in the Gaussian ensembles in the infinite-dimensional case, in which the orthogonal ensemble underlies the unitary and symplectic ensembles.

Next-nearest-neighbor spacing distributions have been obtained for the Gaussian orthogonal and unitary ensembles in three dimensions, showing no appreciable difference in the range of major probability from results in very large dimension [Porter (*57*); Kahn and Porter (*58*)]. Use of generalized ensembles based on the classical polynomials affects the eigenvalue distribution (hence the level density). For example, the nearest-neighbor spacing distribution obtained in the unitary ensemble based on the Legendre polynomials is identical to that of the circular unitary ensemble in the infinite-dimensional case [Leff (*59, 60*)]. In the unitary ensemble the spacing distributions are implied by a certain kernel function [Fox and Kahn (*61*)].

The situation when only energy level positions are known but not the other associated symmetry quantum numbers (like spin and parity) was discussed [Porter and Rosenzweig (*39*); Fox and Kahn (*61*); Gurevich and Pevsher (*62*); Lane (*63*); Trees (*64*)]. The Brownian motion model for the eigenvalue distribution, and spacing distributions, were also discussed [Dresner (*65*); Dyson (*66*); Favro, Kahn, and Mehta (*67*)].

To conclude this chapter we will discuss, following Brody, Flores, French, Mello, Pandey, and Wong (*42*), the spectral rigidity and the separation of fluctuations and secular behavior.

6.6 THE SPECTRAL RIGIDITY

From the discussion presented in Chapter 5 and in this chapter, one infers that fluctuation properties, measured in locally defined units, are the same at both high and low energies, even though the spacing unit varies by a very large factor. (This factor is of the order of ten thousands, for instance, for a heavy nucleus, for which $D \cong 10$ eV in the slow neutron region and $D \cong$ keV in the ground state.) To discuss the region between the two limits, one should have more insight about the way in which the spectrum expands as the energy goes down. In particular, one should ask if there are random aspects to this behavior, or if it is understandable in terms of a small number of significant parameters [Brody *et al.* (*42*)].

A partial answer to this problem emerges from the shell-model calculations, examples of which are given in Figure 6.1, where in each case a shell-model spectrum is compared with a fluctuation-free spectrum which is derived from it and is represented by a small number of parameters. A measure of the long-range rigidity is, therefore, given by the rms deviation, as measured in local spacing units, between the spectrum and its fluctuation-free form. In the 1206-dimensional spectrum shown of ^{24}Mg with $J = 2$, $T = 0$, described via $(ds)^8$, the secular variation, or equivalently, the smoothed spectrum itself, is described satisfactorily by four parameters. It is remarkable that the smoothed spectrum deviates, on the rms average, by only a single spacing unit from its parent. This behavior holds over the entire spectrum. It can also be found in other examples of varying dimensionalities.

In the central region, where the secular variation is negligible, the spectrum shows little deviation from that of a rigid one, the uniformly spaced or "picket-fence" spectrum. Figure 6.1, case c, gives an example of this sort. There seems to be no reason to doubt that this rigidity applies also to physical spectra. Indeed, formal analysis in which one rederives and extends the formalism of Dyson and Mehta (*54*), who first pointed out the spectral rigidity, strongly suggest that this is the case.

In giving these examples, we have not indicated where the "smoothed" or "locally uniform" spectrum comes from yet, since fluctuations are defined in terms of departure from local uniformity. An understanding of this point is quite essential and some simple considerations might make things clearer.

Consider, for instance, the density $\rho(x)$ of a discrete spectrum. Then $\rho(x)$ is proportional to a sum of delta functions,

$$\rho(x) = d^{-1} \sum_r \delta(x - E_r), \tag{6.24a}$$

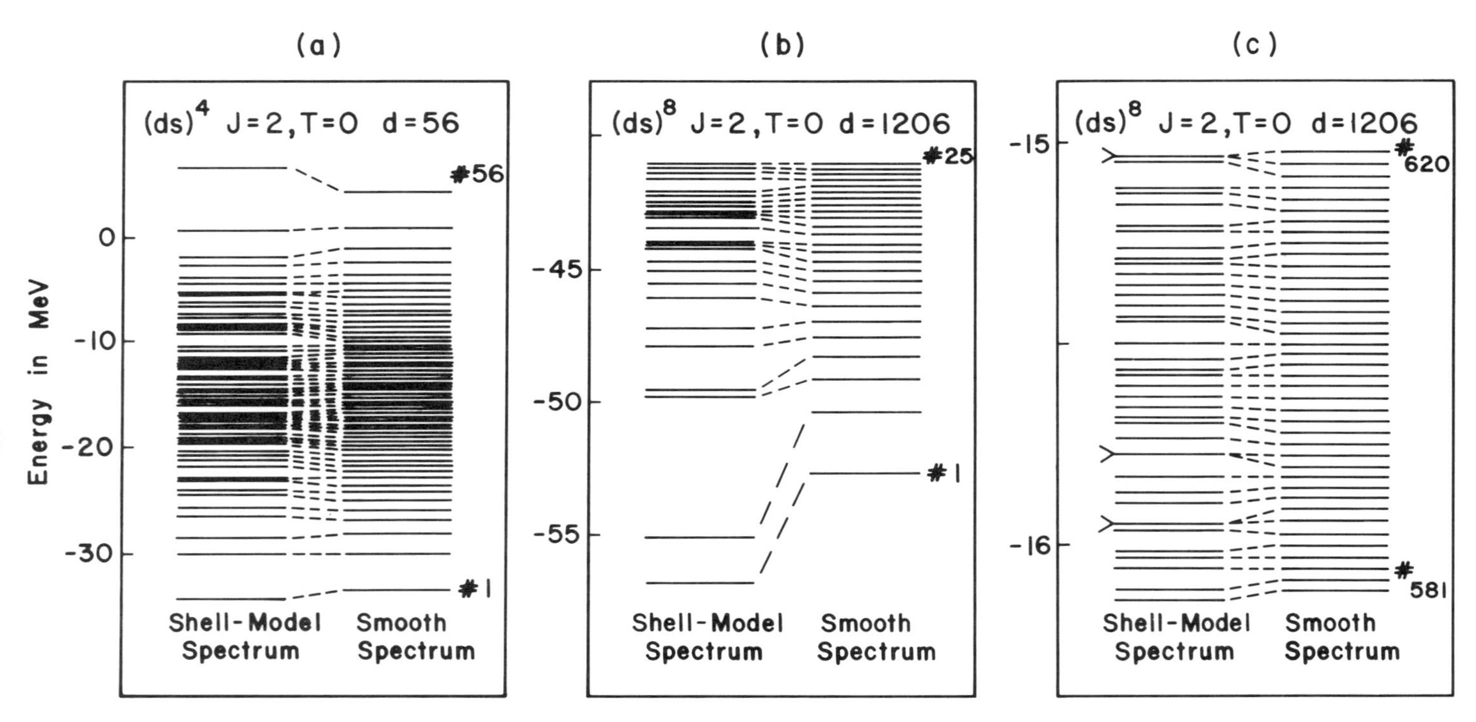

Figure 6.1 Examples of shell-model spectra and the smoothed spectra derived from them by using Eq. (6.27), as illustrated in Fig. 6.2 [Brody *et al.* (*42*)]. The first example shows a complete 56-dimensional spectrum; the other two show small segments of the 1206-dimensional spectrum.

whereas its distribution function (which is a staircase function, as shown in Figure 6.2) is given by

$$F(x) = \int_{-\infty}^{x} \rho(z)\,dz . \tag{6.24b}$$

In Eqs. (6.24), the dimensionality d of the space, namely, the total number of levels considered, was inserted so that the normalization condition

$$\int \rho(x)\,dx = 1 \tag{6.25}$$

will be satisfied, and thus $\rho(x)$ is a probability density. (For an unbounded spectrum, one has to omit the d factor and deal thereby

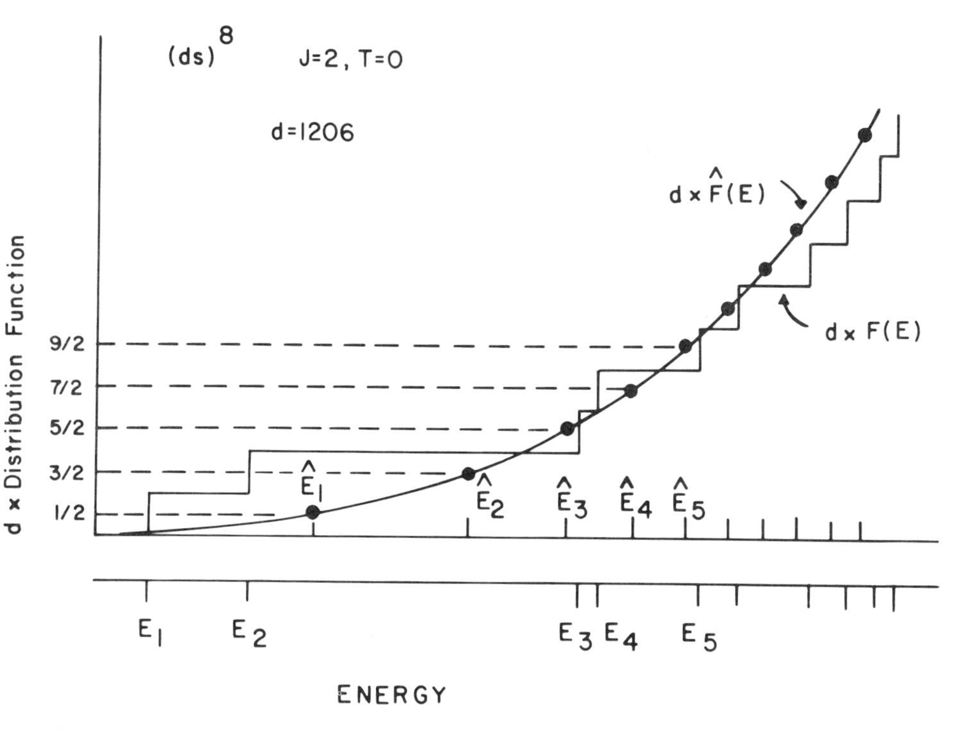

Figure 6.2 An exact distribution function $F(E)$, its smoothed approximation $\hat{F}(E)$, and the spectra E_i and $\hat{E}_i$ which they represent [Brody *et al.* (*42*)]. The example shown is that of Fig. 6.1, case b.

with the actual density of levels.) $F(x)$ then has discontinuities at the eigenvalues,

$$F(E_{r-}) = \frac{r-1}{d}, \tag{6.26a}$$

$$F(E_{r+}) = \frac{r}{d}, \tag{6.26b}$$

where $r = 1, 2, \ldots, d$.

If we have a smoothed (namely, continuous) approximation to the distribution function $F(x)$, let us say $\hat{F}(x)$, the corresponding spectrum would then be defined as the set of values $\hat{E}_r$ satisfying

$$\hat{F}(\hat{E}_r) = \frac{r - \frac{1}{2}}{d}. \tag{6.27}$$

The level deviation is then given by $(E_r - \hat{E}_r)/D$, where D is the local average spacing.

The problem is, then, how to carry out a proper smoothing, whereby $F(x) \to \hat{F}(x)$, of a given spectrum. Empirical methods of doing this for studying fluctuations via a running average, for instance, are usually without a significant physically or mathematically basis.

Broadly speaking, the situation is as follows: There must be some natural limiting distribution (the same for a wide class of systems) and a corresponding expansion of the distribution function in terms of components of varying "wavelengths" (the longest of the order of the spectrum span and the shortest of the order of the mean spacing) built upon the limiting distribution.

If we order the components, which can usefully be called "excitations," according to their "wave number" by a parameter ζ (= 1, 2, 3, . . .), the secular behavior will then be given by truncating the expansion at some values $\hat{\zeta}$ of ζ. Such an operation then introduces a local smoothing. In doing this we are, of course, tacitly assuming that there is a real separation between the secular behavior and the fluctuations. In actual cases, this assumption should be verified.

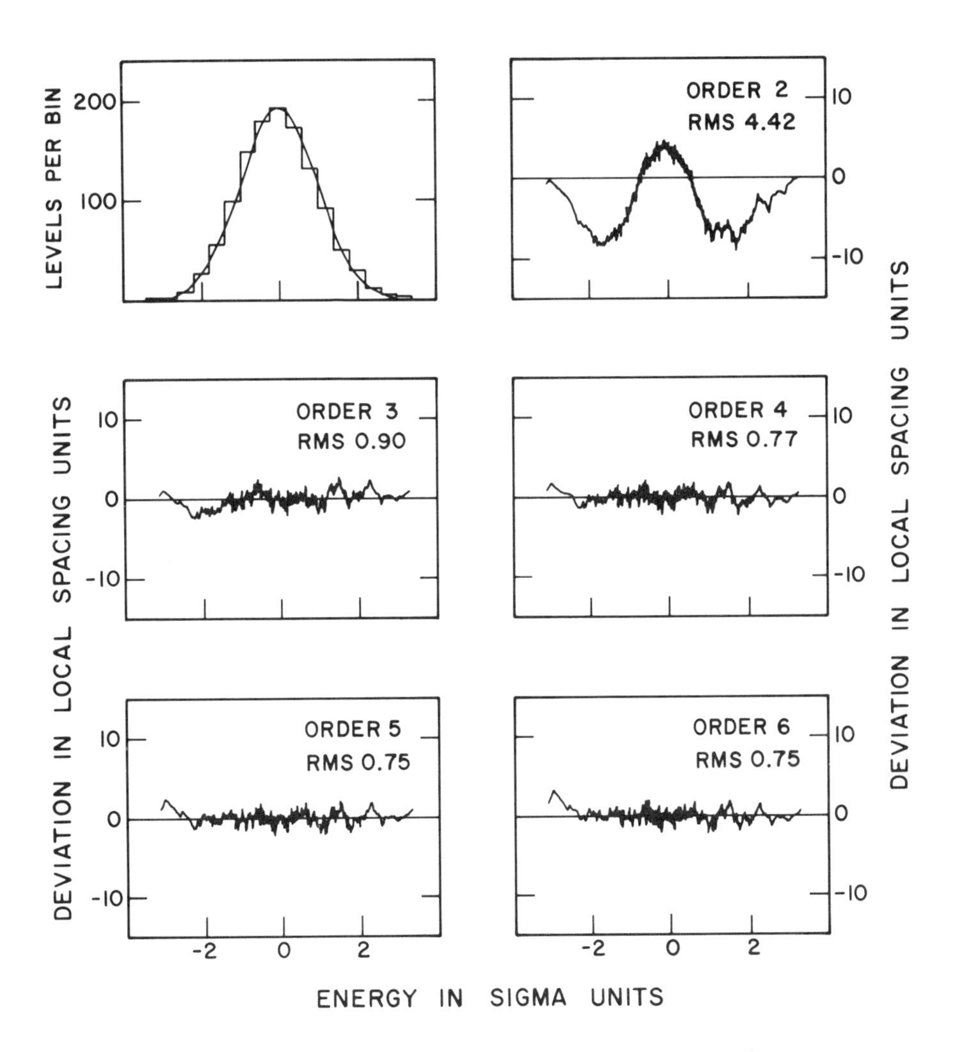

Figure 6.3 Partial normal-mode decomposition of a shell-model spectrum [Brody *et al.* (*42*)]. The spectrum is for the 1206-dimensional $(ds)^8$ with $J = 2$, $T = 0$. The spectral histogram in the upper-left figure shows that the distribution is essentially Gaussian. The level-to-level deviations between the true spectrum and its Gaussian version are displayed in the second figure and are large.

An instructive example of this sort is given in Figure 6.3, which deals once again with the 1206-dimensional shell-model spectrum. The figure also gives the smoothed spectra derived from the shell-model spectrum, and the residual level-to-level fluctuations. The first inset figure gives a simple histogram of the spectrum, along with the corresponding Gaussian density.

In the other figures, which show the level-to-level deviations, the cut-off parameter $\hat{\zeta}$ is varied between 2 (which defines a Gaussian distribution) and 6. A very rapid convergence is seen, and the "final" rms value for the deviation ($\cong$ 0.75 local spacings) is attained with $\hat{\zeta} = 4$. Moreover, it is clear from these figures that in order to reduce the deviation significantly, extremely high-order excitations of the Gaussian density would be needed. Such excitations are of order which is comparable with the dimensionality or, in other words, of wavelength comparable with the spacing.

It thus appears that the fluctuations in the spectrum discussed in Figure 6.3 have an existence which is independent of the secular behavior. This result is completely in line with the results obtained from the slow-neutron and the nuclear-table data, according to which, at least, the nearest-neighbor spacing distributions remain unaltered, even when the spacing parameter varies over a range of ten thousand.

The above conclusion is also supported by the results for other large spectra which show the same rapid convergence of the density to within fluctuations. For example, we find for the three 839-dimensional $(ds)^{12}$ spectra (^{28}Si, $J = T = 0$, with three different effective interactions) that, with $\hat{\zeta} = 6$, the rms deviations are 0.77, 0.79, and 0.82 mean local spacings. It thus appears that the fluctuations for pure sequences always display the same pattern.

In the next chapter, the problem of the distribution of the widths will be discussed.

CHAPTER 7
DISTRIBUTION OF THE WIDTHS

7.1 WIDTHS OF ENERGY LEVELS

We now discuss the fluctuation properties of the *widths* of energy levels, that is, of the square of the wave function at the nuclear or atom surface, and the expectation values. Although a detailed theoretical argument is yet lacking, Scott (*68*) and Porter and Thomas (*69*) found, on the basis of experimental results by Hughes and Harvey (*70, 71*), that the probability that the value of the wave function is between γ and $\gamma + d\gamma$ is

$$\frac{1}{\sqrt{2\pi\overline{\gamma}^2}} e^{-\gamma^2/2\overline{\gamma}^2} \, d\gamma. \tag{7.1}$$

Here $\overline{\gamma}^2$ denotes the average value of γ^2. Equation (7.1) essentially shows that the matrix elements in complex spectra produce a Gaussian distribution, and represents a generally accepted rule, the so-called Porter-Thomas distribution. It is also well confirmed experimentally.

The calculation of the statistics of the matrix elements, in general, involves two self-adjoint operators, the Hamiltonian H defining the coordinate system, and another operator M representing the physical quantity such as the dipole moment, whose matrix elements one wishes to find [Wigner (*6*)].

Let us assume that the two operators H and M are both real. This means there is a coordinate system in which all permissible operators, representing the physical quantity one is interested in, have real matrix elements. If one chooses the coordinates in such a way that the states represented by them are time-inversion invariant, then the matrix elements of the Hamiltonian become real. If the physical quantity one is interested in is time-inversion invariant, then the matrix elements of its operator will also be real.

Most interesting physical quantities do have the property of time-inversion invariance, i.e., retaining their signs when the directions of the velocities are reversed, or the opposite property of reversing their signs if the directions of velocities are reversed. The operators representing these physical quantities, such as the magnetic moment, that reverse their signs when the velocities are reversed are Hermitian and purely imaginary, hence skew-symmetric.

The calculation of the distribution function of the matrix elements of time-reversal-invariant operators can be modified and applied to antiinvariant operators. Therefore the same distribution function is expected for the matrix elements of the antiinvariant operators as for real, time-inversion-invariant operators.

It is also assumed that the density of the characteristic values of the physical quantity is an even function of the characteristic value. This condition is fulfilled for the antiinvariant operators, but also all known time-inversion-invariant transition operators satisfy this condition.

7.2 THE PORTER-THOMAS DISTRIBUTION

Let, then, use be made of a coordinate system whose axes are the characteristic functions of the operator M (and not of the Hamiltonian

H) and let the characteristic values be denoted by μ. The matrix element can then be given by $m = \Sigma\, \mu_i x_i y_i$, where x_i and y_i are the coordinates of the states between which the matrix element is taken. Furthermore, x_i, y_i, and m are real. The problem is then to calculate the distribution of m for the Hamiltonians of the ensemble chosen. If one assumes that this is rotationally invariant in the Hilbert space, then one can calculate the distribution of the above expression for m when the vectors x and y are each normalized to unitary but remain perpendicular to each other,

$$\Sigma\, x_i^2 = \Sigma\, y_i^2 = 1, \quad \Sigma\, x_i y_i = 0.$$

This calculation can be carried out in detail, leading to the fact that the distribution is Gaussian:

$$\frac{1}{\sqrt{2\pi\bar{\mu}^2}}\, e^{-m^2/2\bar{\mu}^2}. \tag{7.2}$$

Introducing the transition rate $\Gamma = m^2$, one then obtains from Eq. (7.2) the distribution

$$\frac{1}{\sqrt{2\pi\bar{\mu}^2\Gamma}}\, e^{-\Gamma/2\bar{\mu}^2}. \tag{7.3}$$

Equation (7.3) is the familiar Porter-Thomas distribution mentioned at the beginning of this chapter. It is confirmed experimentally in the case of neutron emission and various moments in atomic physics [Porter (*72*)]. [For more on the expectation value problem, the reader is referred to Ullah and Porter (*73, 74*).]

It is worthwhile to comment on the assumptions that led to the distribution formula (7.3). It has been shown by Rosenzweig (*75*) that it is plausible that the replacement of Hilbert space by a space of high dimensionality is justified here more than in previous cases discussed. The assumption that the matrices considered were real is also important. An ensemble of complex Hermitian matrices would lead to a transition probability Γ (proportional to the absolute square of the matrix element), $(\bar{\mu}^2)^{-1}\exp(-\Gamma/2\bar{\mu}^2)$, which is different from Eq. (7.3). As Wigner has pointed out, this is significant.

There are experiments which show that the actual Hamiltonian is not strictly time-inversion invariant but has a small antiinvariant part. This part would manifest itself in the region where the levels are close to each other.

Hence, an experimental check on the Porter-Thomas distribution (see Section 7.4) would give an indication on the magnitude of the non-time-inversion-invariant part of the Hamiltonian. The effect here is similar to what would occur in the level-spacing distribution; the non-time-inversion-invariant part of the Hamiltonian would manifest itself in this case in an added repulsion of the levels.

In the rest of this chapter we will discuss, following Brody, Flores, French, Mello, Pandey, and Wong (*42*), strength fluctuations and the Porter-Thomas distribution.

7.3 STRENGTH FLUCTUATIONS AND COLLECTIVE BEHAVIOR

It is worthwhile mentioning that the Porter-Thomas description of the transition amplitudes is, more or less, valid down to the ground-state domain (in correspondence, therefore, with the behavior of the energy-level fluctuations). This observation is usually supported by the inspection of a limited number of shell-model calculations, and by an analysis of some other data. It could also be predicted on the basis of ensemble calculations, when the presence of strong collectivities is not considered.

Let us now consider not the individual strengths, but rather the sum of the strengths which originate with a given starting state. We then ask, how this sum varies as the starting state varies over an energy band. If the strength from every starting state splits into ν equal parts, we then obtain a $\chi_\nu{}^2$ distribution, whose centroid $\mathcal{E}_s$ and variance σ_s^2 are related by

$$\sigma_s^2 = \frac{2\mathcal{E}_s^2}{\nu}.$$

It is easy to extend this result so as to incorporate the secular variation of strength sum with starting-state energy, and to take into account the fact that the strength from a given state does not split evenly among the final states. One then finds the effective number of accessible states (the analog of the effective number of open channels in reaction theory) to be given by [Brody *et al.* (*42*)]:

$$\hat{\nu} = \frac{2\hat{\mathcal{E}}_s^2}{\hat{\sigma}_s^2}, \tag{7.4}$$

where $\hat{\mathcal{E}}_s$ and $\hat{\sigma}_s^2$ are certain local averages of the strength centroids and variances. A small value of $\hat{\nu}$, say $\hat{\nu} \cong 1$, then describes a situation in which the strength originating in the starting region is very little fragmented, most of it going to a single final state.

There are two different cases here: The states in the starting region may have comparable total strengths (but all unfragmented); or there may be a dominant starting state, from which originates most of the strength of the region considered. In the latter, one finds a single, very strong, transition; while in the former, there may be many fairly strong ones. However, each case would be recognized as displaying a strong collectivity.

Given now the parameters of the model spaces and Hamiltonians, the next step is in the actual evaluation of $\hat{\mathcal{E}}_s$ and $\hat{\sigma}_s^2$. This depends, of course, on the smoothed strength and the smoothed square of the strength. Some methods were developed and tested for calculating these quantities in terms of certain traces (or integrals) which can be evaluated directly from the parameters of the model space and the Hamiltonian. Using Eq. (7.4), they then yield an explicit statistical method for predicting certain types of collective behavior.

It is not obvious whether Eq. (7.4), in general, is useful for predicting collectivity. The formula is derived, after all, from the Porter-Thomas law, which we might expect to be modified by the existence of strong collectivities. It is not, then, *a priori* obvious that its application in such cases is really self-consistent.

On the other hand, the only Porter-Thomas feature which is actually used is Eq. (7.4) describing the relationship between the centroid and the variance of the strength distribution. In addition, one has an excellent agreement for the shell-model cases.

Accordingly, it does seem promising that, not only do fluctuations extend to the ground state, but they carry information about phenomena which are usually regarded as highly "non-statistical." Somewhat similar relationships are found between statistical behavior and symmetries.

A general conclusion which we may draw is that we must not separate phenomena in complicated systems into a nonstatistical part, related to "real" physics, and another part, connected with statistical behavior. In some cases, this separation might be useful. However, in general, things are both more complicated and a great deal more interesting to be separated. One can then study a lot from the two kinds of physical behavior, and, in particular, from the relationship between them.

7.4 THE PORTER-THOMAS DISTRIBUTION AND EXPERIMENTAL DATA

The main question now is whether or not locally renormalized transition amplitudes behave as independent Gaussian random variables [Brody *et al.* (*42*)]. For transitions between the states of a given system, defined by a specific Hamiltonian, we rely on the ergodic behavior (see Appendix C) in making a comparison between theoretical ensemble results and experimental spectral results.

Of course, there is also the possibility of combining results for several systems which may extend, for instance, across the nuclear table. In analyzing this question, one cannot attempt a detailed review of the data. One would rather examine, instead, only a few of the especially relevant experiments and computer calculations.

The first question is whether or not the locally renormalized strengths have a χ_1^2 (Porter-Thomas) distribution. The experience here shows that the width distribution, e.g., that of Γ_n in (n, γ) reactions, is indeed Porter-Thomas. This is so at least as long as no doorway states or other intermediate-state structures are present. In fact, many statistical tests of the experimental data (for missing levels and for partial identification of spin), are based on the assumption that this is the distribution.

As an example with very high-quality data is that of the neutron widths for $^{166}\mathrm{Er} + n$, 174 widths with neutron energy $\leqslant 9.5$ keV. Fig. 7.1 shows a histogram of the data drawn to logarithmic scale, and the corresponding Porter-Thomas distribution. As is seen, the

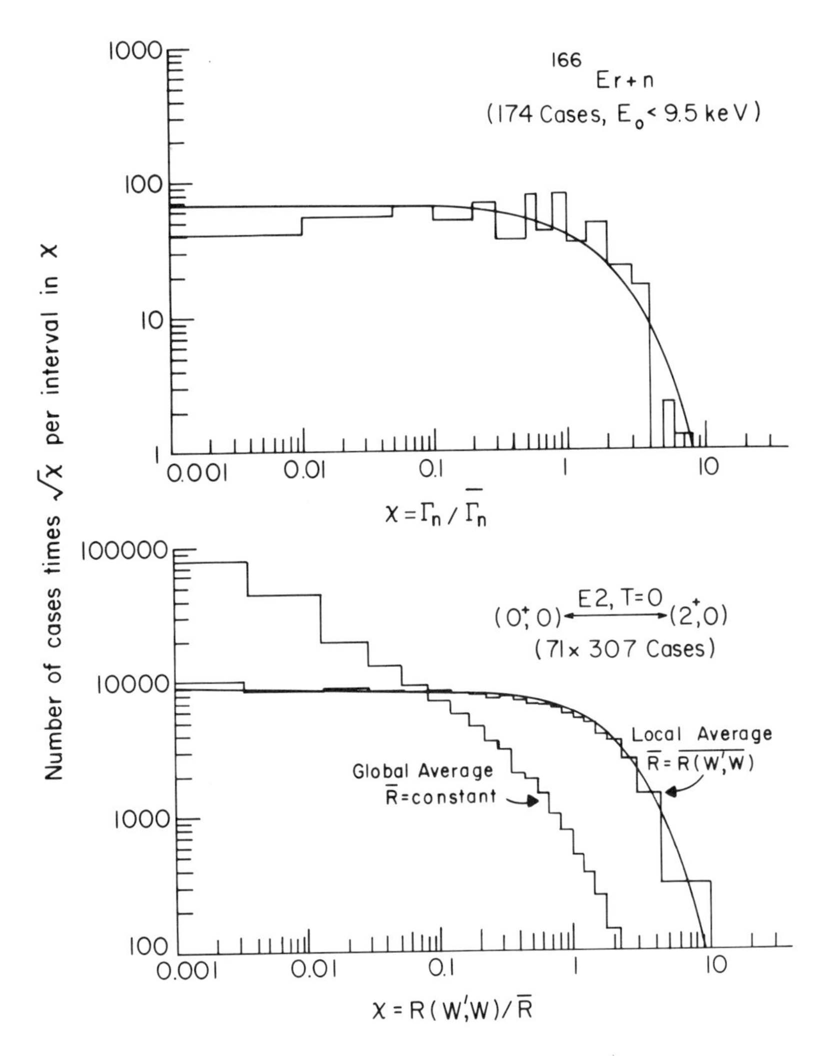

Figure 7.1 Histogram of the reduced neutron widths for ^{166}Er + n is compared with the Porter-Thomas distribution [Brody *et al.* (*42*)]. Good agreement is seen. A similar comparison is made for $E2$ strengths for transitions between two shell-model spaces $(ds)^6$, $T = 0$, $J = 0, 2$. When the strengths are locally renormalized, as is suggested by the theory, the agreement is again good. However, when this renormalization is not carried out, the distribution has a radically different shape.

fit is quite good. Other plots of this sort (for neutron width data of 144 widths of even targets), show the same comparison, with about the same quality of agreement. Once again, it is clear that the Porter-Thomas distribution is a good representation of the data.

Further examinations can be made by calculating a χ^2 confidence limit, or by using a maximum-likelihood procedure based on the number of degrees of freedom [Porter and Thomas (*69*)]. In the latter procedure (which is more commonly used), one calculates ν, the number of degrees of freedom, assuming that the distribution itself is χ^2_ν, with ν not necessarily an integer. A statistically significant departure from $\nu = 1$ is then an indication of a failure of the Porter-Thomas assumption.

However, there is a tendency for the data to have systematic errors which almost always result in larger calculated ν values. For example, $\nu = 1.20^{+0.02}_{-0.15}$ was found for 12 final states, and 23 *s*-wave resonances with $E_n < 600$ keV in $^{238}\mathrm{U}(n, \gamma)^{239}\mathrm{U}$; whereas the systematic error due, for example, to unresolved components in the γ-ray peaks, was estimated to increase ν by as much as 20%.

Accordingly, it is probable that many other known $\nu > 1$ (n, γ) results may be due to the same cause. With some simple statistical assumptions on the errors, unbiased estimates of ν were given and a good agreement with the Porter-Thomas distribution for M1 transition strengths in the $^{105}\mathrm{Pd}\ (n, \gamma)$ reaction was found. For $E1$ transitions, one finds ν values slightly larger than unity. This may again be due to an inadequate accounting of the errors.

More examples, like the γ-rays from nine 3^- and seven 4^- resonant states to a total of 41 final 2^+, 3^+, 4^+ states below 2600 keV in the $^{149}\mathrm{Sm}(n, \gamma)$ reaction, were analyzed. It was found that, for transitions to the eight 2^+ final states, there is no significant departure from the Porter-Thomas distribution.

However; whereas the maximum-likelihood tests for the thirty three 3^+ and 4^+ final states give results which are consistent with the Porter-Thomas distribution, the same tests for the lowest thirteen 3^+ and 4^+ states (below 2196 keV) give a small departure. The detailed analysis of the probable errors seems to indicate that the deviation from the Porter-Thomas distribution is a genuine one.

The effects of the secular variation in the strength are best studied by means of shell-model calculations, for which they are large. In Fig. 7.1 also is given the $E2$ strength distribution derived from the $\cong$ 20,000 transitions between the $J = 2$, $T = 0$ and $J = 0$, $T = 0$ $(ds)^6$ shell-model states (constructed for a "realistic" Hamiltonian). The strength distribution is seen to be quite different from the Porter-Thomas distribution.

On the other hand, when the strengths are locally renormalized (to eliminate the secular variation) the distribution, given also in Fig. 7.1, agrees beautifully with the Porter-Thomas distribution. Similar results hold for M1 and E4 transitions in $(ds)^6$, and for distributions in which the strengths are taken from restricted energy domains (which would, when sufficiently restricted, yield the Porter-Thomas distribution without renormalization).

The above described shell-model calculations give an excellent further demonstration that the fluctuations, when properly measured, are invariant under mappings and embeddings of the Hamiltonian.

We now turn to the question of strength correlations in cases where the ordinary Porter-Thomas distribution should be applicable. We consider, in particular, the case of "resonance" correlations in which a "statistical" state (a member of a sequence) makes transitions to two channel states. One can then show that the correlation coefficient, in this case, is not necessarily zero, but it is non-negative for the ordinary statistical model.

Some experiments seem often to indicate on significant correlations, which in some cases disappeared with better γ-ray or neutron energy resolution. Careful studies of the reaction $^{238}U(n, \gamma)$, for instance, referred to above, have shown, except in one case, no significant correlations. The γ transitions to 15 final states originating from each of the 23 s-wave resonances with $E_n < 600$ keV were examined. For each of the 105 γ-ray final-state pairs one can, then, determine a correlation coefficient from the series of resonances.

A question which consequently arises in the analysis is whether or not the observed correlations may arise from finite-sample effects which are large because the resonance series is short. This question was studied by comparing the histograms of the correlations with that arising from a large number of similar examples constructed via Monte-Carlo calculations (which make use of a set of uncorrelated χ_1^2 strength distributions). The conclusion was that the one very large correlation coefficient (+0.81 between γ rays with energies

3991 and 3982 keV), which is observed, cannot for certain be a real effect. So is the case for correlations between the neutron and the γ-ray widths.

On the other hand, it was found that in the ^{169}Tm(n, γ) reaction, there is a large correlation (+0.274) between the neutron and the γ-ray widths, and a smaller average γ-γ correlation (+0.088). This indicates that the excited states have a larger-than-statistical component of a single-particle excitation of the ground state.

It should be pointed out that in both of these experiments, the correlations are "initial state" or "resonance" ones, which involve a sequence of such states. These are in contrast to "final-state" correlations, which result from averaging over the final states for any two given initial states. Many examples of large correlations of both types were given by Lane. In fact, there is a tendency for a given target and neutron partial wave to show both or neither.

Finally, we mention that transition-amplitude correlations were also determined for proton resonances, and in particular on ^{48}Ti. The relation between the strength and amplitude correlations was found to be not satisfied for much of the data. No satisfactory explanation has yet been found for this anomaly.

In the next chapter, the relationship between the symplectic group and the quaternions will be discussed in detail.

CHAPTER 8
SYMPLECTIC GROUP AND QUATERNIONS

8.1 THE SYMPLECTIC GROUP

A set of matrices B form an N-dimensional symplectic group, denoted by $\mathrm{Sp}(N)$, if B is a $2N \times 2N$ unitary matrix and satisfies the relation [see Eq. (3.10), and Dyson (*4*); Weyl (*33*)]:

$$Z = BZB^t . \tag{8.1}$$

[The matrix Z was defined in the text and is given by Eq. (3.12).] It is well known [Chevalley (*32*)] that the algebra of the symplectic group can be expressed most naturally in terms of quaternions. We here give a brief discussion of quaternions.

8.2 QUATERNIONS

Introducing the standard quaternion notation for 2×2 matrices

$$\tau^1 = \begin{bmatrix} 0 & -i \\ -i & 0 \end{bmatrix}, \quad \tau^2 = \begin{bmatrix} 0 & -1 \\ 1 & 0 \end{bmatrix}, \quad \tau^3 = \begin{bmatrix} -i & 0 \\ 0 & i \end{bmatrix}, \tag{8.2}$$

satisfying the multiplication table

$$(\tau^1)^2 = (\tau^2)^2 = (\tau^3)^2 = -1, \tag{8.3}$$

$$\left\{\tau^i, \tau^j\right\} = \epsilon^{ijk}\tau_k, \tag{8.4}$$

where $\left\{\tau^i, \tau^k\right\}$ denotes an anticommutator $\tau^i\tau^k + \tau^k\tau^i$ and ϵ^{ijk} is the skew-symmetric tensor such that $\epsilon^{123} = 1$. All $2N \times 2N$ matrices will be considered as cut into N^2 blocks of 2×2, and each 2×2 block is regarded as a quaternion.

Hence a $2N \times 2N$ matrix with complex elements becomes an $N \times N$ matrix with complex quaternion elements. For example, the matrix Z can now be written as

$$Z = \tau^2 I, \tag{8.5}$$

where I is the $N \times N$ unit matrix. It is interesting to note that the matrix rules for multiplication are not changed by this transcription.

A quaternion g is called *real* if it has the form

$$q = q^0 + (\mathbf{q} \cdot \tau), \tag{8.6}$$

and the coefficients q^0, q^1, q^2, and q^3 are real numbers. Hence a real quaternion is not a real 2×2 matrix. The *conjugate* quaternion q^* to the complex quaternion q is defined by

$$q^* = q^0 - (\mathbf{q} \cdot \tau). \tag{8.7}$$

Note that the conjugate quaternion q^* is *different* from the *complex conjugate* quaternion $\bar{q}$ defined by

$$\bar{q} = \bar{q}^0 + (\bar{\mathbf{q}} \cdot \tau). \tag{8.8}$$

A quaternion q satisfying $\bar{q} = q$ is real. A quaternion q with $q^* = q$ is a *scalar*. The *Hermitian conjugate* quaternion $q^\dagger$ is defined by applying both types of conjugation together,

$$q^\dagger = \bar{q}^* = \bar{q}^0 - (\bar{\mathrm{q}} \cdot \tau). \tag{8.9}$$

8.3 MATRICES AND QUATERNIONS

Consider now a general $2N \times 2N$ matrix A which one writes as an $N \times N$ matrix Q with quaternion elements q_{ij}, with $i, j = 1, 2, \ldots, N$. The standard matrix operations on A then reflect themselves on Q as follows:

Transposition

$$(Q^\dagger)_{ij} = - \tau^2 q_{ji}^* \tau^2 \; ; \tag{8.10}$$

Hermitian conjugation

$$(Q^\dagger)_{ij} = q_{ji}^\dagger \; ; \tag{8.11}$$

Time reversal

$$(Q^R)_{ij} = q_{ji}^* . \tag{8.12}$$

The usefulness of the quaternion algebra is a result of the simplicity of Eqs. (8.11) and (8.12). Using Eqs. (8.11) and (8.12), one sees that the condition

$$Q^R = Q^\dagger \tag{8.13}$$

is necessary and sufficient for the elements of Q to be real quaternions. When Eq. (8.13) is satisfied one calls Q *quaternion real.*

A unitary matrix B satisfying Eq. (8.1) is automatically quaternion real, and satisfies the conditions

$$B^R = B^\dagger = B^{-1}, \tag{8.14}$$

which can be considered as defining the symplectic group. Matrices S representing physical systems are not quaternion real. They are unitary and self-dual, i.e.,

$$S^R = S, \quad S^\dagger = S^{-1}. \tag{8.15}$$

8.4 QUATERNION ALGEBRA

The theorem of quaternion algebra states:

Theorem 8.1. *Let H be any Hermitian, quaternion-real, $N \times N$ matrix. Then there exists a symplectic matrix B such that*

$$H = B^{-1} DB, \tag{8.16}$$

where D is diagonal, real, and scalar. The statement that D is scalar means that it consists of N blocks of the form

$$\begin{bmatrix} D_j & 0 \\ 0 & D_j \end{bmatrix}. \tag{8.17}$$

Hence the eigenvalues of H consist of N equal pairs.

The Hamiltonian of any system that is invariant under time reversal and has odd spin satisfies the condition of the theorem stated above. The energy levels of such a system, as was pointed out in Chapter 3, must be doubly degenerate. This is the Kramers degeneracy [Kramers (*28*), Tinkham (*29*)] and the above theorem shows how it appears naturally in the quaternion language.

An extension of the above theorem states that if S_1 and S_2 are two commuting, Hermitian, quaternion-real matrices, then there exists a symplectic matrix B such that

$$S_1 = B^{-1} D_1 B, \qquad S_2 = B^{-1} D_2 B, \tag{8.18}$$

where D_1 and D_2 are diagonal, real, and scalar. From this extension one can prove:

Theorem 8.2. *Let S be any unitary, self-dual, $N \times N$ quaternion matrix. Then there exists a symplectic matrix B such that*

$$S = B^{-1} EB, \tag{8.19}$$

where E is diagonal and scalar. The diagonal elements of E are N complex numbers $\exp(i\theta_k)$ on the unit circle, each being repeated twice.

To prove the theorem, one writes

$$S = S_1 + S_2, \tag{8.20}$$

where S_1 and S_2 are quaternion real. Since the operation of the time reversal applied to a matrix does not involve complex conjugation, when S is self-dual each of S_1 and S_2 must be separately self-dual. Hence S_1 and S_2 are also Hermitian. Moreover, since S is unitary,

$$S^\dagger S = (S_1 - iS_2)(S_1 + iS_2) = 1. \tag{8.21}$$

Separating the quaternion-real and -imaginary parts of Eq. (8.21), one finds

$$S_1{}^2 + S_2{}^2 = 1, \quad S_1 S_2 - S_2 S_1 = 0. \tag{8.22}$$

Hence S_1 and S_2 commute, and therefore the extension of Theorem 8.1 applies. If B is chosen to satisfy Eq. (8.18), then Eq. (8.19) will hold with

$$E = D_1 + iD_2, \tag{8.23}$$

being diagonal and scaler. Let d_k and d_k' be the corresponding eigenvalues of D_1 and D_2. Equation (8.22) then gives

$$(d_k)^2 + (d_k')^2 = 1. \tag{8.24}$$

This shows that d_k and d_k' can be taken as

$$d_k = \cos\theta_k, \quad d_k' = \sin\theta_k, \tag{8.25}$$

and therefore the diagonal elements of E become

$$e_k = d_k + id_k' = \exp(i\theta_k), \tag{8.26}$$

each occurring twice.

The analog of Theorem 8.2 for the even-spin case is:

Theorem 8.3. *Let S be any unitary, symmetric, $N \times N$ matrix. Then there exists a real orthogonal matrix R such that*

$$S = R^{-1} ER, \tag{8.27}$$

where E is diagonal. The diagonal elements of E are N complex numbers $\exp(i\theta_k)$ *on the unit circle.*

The proof of this theorem is the same as that for Theorem 8.2 if one replaces self-dual for symmetric, quaternion-real for real, and symplectic for orthogonal.

8.5 SYMPLECTIC ENSEMBLE

Returning now to the odd-spin case and defining the *symplectic ensemble* E_4, the odd-spin analog of the orthogonal ensemble E_1. One then works in the space T_4 of unitary self-dual quaternion matrices. We now define an *invariant measure* in T_4 in spite of the fact that the matrices of T_4 do not form a group.

First one notices that every matrix S in T_4 can be written in the form

$$S = U^R U, \tag{8.28}$$

where U is unitary. Given S, the unitary matrix U has a freedom of

$$U \to BU, \tag{8.29}$$

where B is an arbitrary symplectic matrix.

An infinitesimal neighborhood of S in T_4 is given by

$$S + dS = U^R \, [1 + i\, dM] \, U, \tag{8.30}$$

where dM is a quaternion-real, self-dual, infinitesimal matrix with elements

$$dM_{ij} = dM_{ij}^0 + (d\mathbf{M}_{ij} \cdot \boldsymbol{\tau}), \tag{8.31}$$

where the real coefficients dM_{ij} satisfy

$$dM^0_{ij} = dM^0_{ji}, \quad dM^\alpha_{ij} = - dM^\alpha_{ji}, \quad \alpha = 1, 2, 3. \tag{8.32}$$

One has $N(2N - 1)$ independent real variables dM^α_{ij}, which vary through some small intervals of length $d\mu^\alpha_{ij}$. The neighborhood of S so defined therefore has the measure

$$\mu(dS) = \prod_{\alpha,i,j} d\mu^\alpha_{ij}. \tag{8.33}$$

The symplectic ensemble E_4 is therefore defined like E_1, with the statistical weight of the neighborhood dS in E_4 given by

$$P(dS) = (V_4)^{-1} \mu(dS), \tag{8.34}$$

where V_4 is the total volume of the space T_4.

Finally one has:

Theorem 8.4. *The symplectic ensemble E_4 is uniquely defined, in the space T_4 of self-dual, unitary quaternion matrices, by the property of being invariant under every automorphism $S \to W^R S W$ of T_4 into itself, where W is a unitary matrix.*

This theorem shows that the symplectic ensemble uniquely represents the notion of uniform a priori probability in the space T_4.

The next chapter will be devoted to a detailed discussion on the Gaussian ensemble.

CHAPTER 9
MORE ON THE GAUSSIAN ENSEMBLE

9.1 THE GAUSSIAN ENSEMBLE

In Chapter 4 we discussed the Gaussian and orthogonal ensembles. We here give some more details [Porter (*19*)].

The Gaussian ensembles [Wigner (*35-38*); Porter and Rosenzweig (*39*); Wigner (*48*)] have the unifying feature that eigenvector and eigenvalue distributions are produced directly from a single set of hypotheses concerning the Hamiltonian of the system which usually reflects the actual dynamics. This means that data on *level widths, expectation values*, and *level spacings* can be used to test the same ensemble.

Let us examine the orthogonal case in which the Hamiltonian can be made a real symmetric matrix. As a model, one considers a real symmetric matrix of dimension N to represent the infinite submatrix of the Hamiltonian H associated with some symmetry property. Hence the Hamiltonian will include $N(N+1)/2$ real parameters.

To associate a measure (volume) with the matrix H, one introduces a metric in the parameter space defined by

$$ds^2 = \mathrm{Tr}(dH\, dH^\dagger), \tag{9.1}$$

where dH represents a matrix whose elements are infinitesimal increments of the elements of the matrix H. Comparing Eq. (9.1) with the standard form of line element

$$ds^2 = \Sigma\, g_{\mu\nu}\, dx^\mu\, dx^\nu, \tag{9.2}$$

one finds that the metric $g_{\mu\nu}$ is diagonal with N of the elements equal to one and the remaining $N(N-1)/2$ elements equal to $\sqrt{2}$. A measure can then be defined as

$$d\mu(H) = 2^{N(N-1)/4}\, dH_{11}\, dH_{22} \ldots dH_{NN}\, dH_{12} \ldots dH_{N-1,N} \tag{9.3}$$

in analogy to the volume element in a Riemannian space

$$dV = (\det g)^{1/2}\, dx_1\, dx_2 \ldots dx_M, \tag{9.4}$$

where M is the dimension of the space, $M = N(N+1)/2$. The quantities $dH_{11}, \ldots, dH_{N-1,N}$ are the elements of the matrix dH.

9.2 THE DISTRIBUTION FUNCTION

With the matrix H one can associate a distribution function

$$P(H) = P(H_{11}, H_{22}, \ldots, H_{NN}, H_{12}, \ldots, H_{N-1,N}), \tag{9.5}$$

such that the differential probability dP that H_{11} is in dH_{11} at H_{11}, etc., is

$$dP = P(H)\, d\mu(H). \tag{9.6}$$

Since the matrix H is Hermitian, and Hermitian matrices do not form a group, one encounters a difficulty in defining a measure in terms of H. However, H is related to a diagonal matrix H_D containing the eigenvalues of H by the orthogonal transformation R,

$$H = RH_D R^{-1} = RH_D R^t. \tag{9.7}$$

The last equation is actually the Schrödinger equation written in matrix form, where R contains the eigenvectors usually written as ψ and H_D contains the eigenvalues.

One can thus associate the volume element (9.3) with R and H_D, where, as is well known, the matrices R do provide a group. Using Eq. (9.7), one can then write

$$dH = R\, dH_D\, R^t + [dR\, R^t, H]. \tag{9.8}$$

Hence the $N(N + 1)/2$ variables of H are transformed into the N eigenvalues in H_D and $N(N - 1)/2$ angles of rotation in the matrix R. One can then insert for dH of Eq. (9.1) its value given in Eq. (9.8) in terms of the eigenvalues in H_D and the angles of R. The measure $d\mu(H_D)$ associated with H_D is given by

$$d\mu(H_D) = dE_1\, dE_2 \ldots dE_N. \tag{9.9}$$

In order to find the measure of R, one writes [von Neumann (*11*)] Eq. (9.7) in terms of matrix elements,

$$H_{\mu\nu} = \sum_{\alpha=1}^{N} E_\alpha R_{\mu\alpha} R_{\nu\alpha}. \tag{9.10}$$

[This approach is somewhat different from that of Hua (*76*). For measure of the group R in three dimensions see Carmeli and Malin (*77*).]

Equation (9.10) defines a transformation from the eigenvalue and eigenvector variables to the Hamiltonian matrix element variables. Hence one can write the measure (9.3) as

$$d\mu(H) = 2^{N(N-1)/4}\, dH_{11} \ldots dH_{NN}, \tag{9.11a}$$

or

$$d\mu(H) = 2^{N(N-1)/4}\, J\, dE_1 \ldots dE_N\, d\alpha_1 \ldots d\alpha_{N(N-1)/2}, \tag{9.11b}$$

where $\alpha_1, \ldots, \alpha_{N(N-1)/2}$ are the $N(N-1)/2$ parameters of the orthogonal matrix R. In Eqs. (9.11) the term J stands for the Jacobian of the transformation (9.10) and is given by the determinant

$$J = \begin{vmatrix} \partial H_{11}/\partial E_1 \ldots \partial H_{11}/\partial E_N & \partial H_{11}/\partial\alpha_1 \ldots \partial H_{11}/\partial\alpha_{N(N-1)/2} \\ \vdots & \\ \partial H_{NN}/\partial E_1 & \ldots \quad \partial H_{NN}/\partial\alpha_{N(N-1)/2} \end{vmatrix}. \tag{9.12}$$

9.3 THE JACOBIAN OF THE TRANSFORMATION

It can be shown that J is a polynomial of degree $N(N-1)/2$ in the eigenvalues. Since the eigenvectors are not unique, if one pair of eigenvalues are equal it follows that the transformation inverse to Eq. (9.10) is singular and therefore J vanishes. Hence J is proportional to

$$\prod_{\lambda<\mu=1} |E_\lambda - E_\mu|. \tag{9.13}$$

Thus J can be written as

$$J = h(\alpha_1, \ldots, \alpha_{N(N-1)/2}) \prod_{\lambda,\mu} |E_\lambda - E_\mu|. \tag{9.14}$$

Since the measure of the n-dimensional orthogonal group can be written as

$$d\mu(R) = h(\alpha_1, \ldots, \alpha_{N(N-1)/2})\, d\alpha_1 \ldots d\alpha_{N(N-1)/2}, \tag{9.15}$$

we see that

$$d\mu(H) = d\mu(H_D)\, d\mu(R) \prod_{\lambda<\mu=1}^{N} |E_\lambda - E_\mu|. \tag{9.16}$$

9.4 SPECIFICATION OF THE DISTRIBUTION FUNCTION

How should the function $P(H)$, appearing in Eq. (9.6), be specified? One knows that it depends only on the eigenvalues and not on the eigenvectors. One requires that $P(H)$ be independent in the matrix elements. Moreover, if one does not know the Hamiltonian H, then one also does not know in what representation H will be diagonal and therefore one does not know R, thus choosing it at random.

Hence it follows that $P(H)$ should be independent of R, namely, $P(H)$ is an invariant function of H. It follows that the requirements of invariance and independence are necessary and sufficient conditions to determine the form of $P(H)$, where, mathematically, invariance mean

$$dP/d\alpha = 0, \tag{9.17}$$

and independence means

$$P(H) = f_{11}(H_{11}) \ldots f_{NN}(H_{NN}) \ldots f_{N-1,N}(H_{N-1,N}). \tag{9.18}$$

A detailed calculation, using Eqs. (9.17) and (9.18), shows that

$$f_{\mu\mu} = \frac{\exp[-(H_{\mu\mu} - E_0)^2/4\alpha^2]}{(4\pi\alpha^2)^{1/2}}; \tag{9.19a}$$

and

$$f_{\mu\nu} = \frac{e^{-H_{\mu\nu}^2/2\alpha^2}}{(2\pi\alpha^2)^{1/2}}; \quad \mu < \nu, \tag{9.19b}$$

with the normalization condition

$$\int_{-\infty}^{\infty} f_{\mu\nu}(H_{\mu\nu})\, dH_{\mu\nu} = 1. \tag{9.20}$$

In these equations a and E_0 are constants. Finally, the expression for the function $P(H)$ becomes

$$P(H) = \frac{e^{-\mathrm{Tr}(H-E_0 I)^2/4a^2}}{(4\pi a^2)^{N(N+1)/4}}, \tag{9.21}$$

normalized such that

$$\int P(H)\, d\mu(H) = 1. \tag{9.22}$$

This is the situation for the orthogonal ensemble.

9.5 THE UNITARY ENSEMBLE CASE

The corresponding formulas in the unitary ensemble are

$$d\mu(H) = d\mu(H_D)\, d\mu(U) \prod_{\lambda<\mu=1}^{N} (E_\lambda - E_\mu)^2, \tag{9.23}$$

where H_D and U are related to H by

$$H = UH_D U^\dagger, \tag{9.24}$$

and

$$P(H) = \frac{e^{-\mathrm{Tr}(H-E_0 I)^2/4a^2}}{(4\pi a^2)^{N^2/2}}. \tag{9.25}$$

In the symplectic case one obtains

$$d\mu(H) = d\mu(H_D)\, d\mu(S) \prod_{\lambda<\mu} (E_\lambda - E_\mu)^4, \tag{9.26}$$

where H_D and S are related to H by

$$H = SH_D S^\dagger, \tag{9.27}$$

and S is symplectic (see Chapter 8), and

$$P(H) = \frac{e^{-\mathrm{Tr}(H-E_0 I)^2/4a^2}}{(4\pi a^2)^{(2N^2 - N)/2}} . \tag{9.28}$$

These results of the Gaussian ensembles can be summarized by writing

$$d\mu(H_\beta) = 2^{\beta N(N-1)/4} \left[\prod_{\mu=1}^{N} dH_{\mu\mu} \right] \left[\prod_{\mu<\nu=1}^{N} \prod_{k=0}^{\beta-1} dH_{k\mu\nu} \right], \tag{9.29}$$

or

$$d\mu(H_\beta) \cong \left[\prod_{\lambda<\mu=1} |E_\lambda - E_\mu|^\beta \right] d\mu(H_D) \times \begin{cases} d\mu(R) \\ d\mu(U) \\ d\mu(S) \end{cases} \tag{9.30}$$

and

$$P(H;\beta) = \frac{e^{-\mathrm{Tr}(H-E_0 I)^2/4a^2}}{(4\pi a^2)^{[N+\beta N(N-1)/2]/2}}, \tag{9.31}$$

with

$$\int P(H;\beta)\, d\mu(H_\beta) = 1. \tag{9.32}$$

Here $\beta = 1$ corresponds to the orthogonal, $\beta = 2$ to the unitary, and $\beta = 4$ to the symplectic ensemble.

In the next chapter a summary of the statistical theory of energy levels and random matrices in physics is given.

CHAPTER 10
SUMMARY

10.1 RANDOM MATRICES AND ENERGY LEVELS

We have seen how the statistical theory of energy levels is related to the theory of random matrices. The theory is applicable to a variety of subjects [see, e.g., Grag (*78*)]: theoretical aspects of resonance reaction theories, statistics of resonance parameters such as level spacings, neutron fission radiative and reaction widths, level densities, fluctuations in cross sections, strength functions and their relation to the optical model, intermediate structure in particle- and photon-induced reactions, and statistical aspects of the decay of the compound nucleus.

However, as Wigner (*79*) has pointed out, the subject is not precisely defined as compared to the theory of statistical mechanics. The latter subject is clearly defined since it deals with time-averaged properties which depend only on the energy and are independent of the other initial conditions, as the quasiergodic theorem asserts.

On the other hand, in the theory of statistical properties of nuclei one considers situations in which one is not interested in as detailed a picture of the nucleus as one is generally interested in physics but tries to find properties and rules which are reasonably simple and general, shared by most nuclei under appropriate conditions.

The role of the Hamiltonian in both statistical mechanics theory and in the statistical theory of energy levels is very important and worth a few remarks [Wigner (*79*)]. It is well known that some of the very interesting laws of statistical mechanics can be derived from the simple assumption that the equations of motion can be derived from a Hamiltonian through the usual variational principle.

10.2 ROLE OF THE HAMILTONIAN

For example, the entropy theorem and the equipartition theorem belong in this category [see, e.g., Joos (*80*)]. Also, much of other work in the theory of statistical mechanics is based on the explicit knowledge of the Hamiltonian itself (which is known in most practical cases). There are, however, no known theorems which are as fundamental as the entropy theorem or the equipartition theorem. In fact, as is well known, one does not even know the nuclear Hamiltonian.

Moreover, some relevant properties of the Hamiltonian are complicated, even though its explicit form is not known. Of course, the Hamiltonian in the statistical theory of energy levels is an operator, and hence a matrix, defined in a certain Hilbert space.

The most natural set of properties to be made use of are those shared by self-adjoint matrices in Hilbert space. This then leads to the concept of ensembles of self-adjoint, or real symmetric, matrices in Hilbert space, which is a definition of the measure for such matrices in Hilbert space. The concept of "vast majority of all self-adjoint matrices" or "practically all self-adjoint matrices" is then mathematically defined.

10.3 THE WISHART DISTRIBUTION

A particular case of distribution, which is of much interest for both physicists and mathematicians, is that of the Wishart distribution. [There seems to be a difference on the use of terminology "Wishart distribution" between physicists and statisticians. See comment by Krishnaiah in (*78*), p. 20.] If one demands that the measure be invariant with respect to unitary transformations, the measure is then an arbitrary function of the invariants of the matrix, multiplied by the differentials of the independent components of the matrix elements.

If, further, one demands that the probabilities of the independent components of the matrix elements be independent of each other, one then obtains, essentially, the Wishart distribution, and the ensemble obtained is such that the number of matrices within unit interval of the independent components of the matrix elements M_{ik} is proportional to $\exp(\alpha \Sigma M_{ii} - \beta \Sigma |M_{ik}|^2)$, where α and β are arbitrary constants. However [Wigner (*79*)], practically none of the matrices of this distribution have characteristics similar to those observed for actual Hamiltonians.

In particular, the density of the characteristic values of most matrices as a function of energy (the semicircle law) has, in the neighborhood of the lower bound, a negative second derivative, whereas the measured density has a positive second derivative with respect to the energy. Hence one should be careful to draw other consequences that are obtained from the Wishart model.

10.4 RECENT STUDIES

To overcome this difficulty, one might use the independent particle model [French and Wong (*81*); Bohigas and Flores (*82*); French and Chang (*83*); Bohigas and Flores (*84*)]. The result obtained by these recent studies is that the density of the levels as a function of the energy is a Gaussian, rather than the semicircle of the Wishart and related ensembles, and at its lowest range its second derivative is positive. Another result [Bohigas and Flores (*82, 84*)] is that the distance of the second and further neighbors is subject to much larger fluctuations than in the Wishart ensemble case. However, the independent particle model has its own restricted validity [Wigner (*79*)].

Another model that might replace the Wishart ensemble is that in which the matrix elements are viewed as particles in Brownian motion. It then follows that the characteristic values can also be viewed as Brownian particles.

A third model is based on an ensemble of matrices of the form $H = m^{\dagger} m + m^{t} m$, where m is again an arbitrary complex matrix. This model seems to promise the important factor of mathematical simplicity [Wigner (*79*)].

[Professor Wigner has kindly informed me (private communication) that his proposal to use $m^{\dagger} m + m^{t} m$, with m a complex Wishartian ensemble member, has proved to be invalid; there is a finite gap in the distribution of the roots at zero and it is at least unlikely that there is an energy region with a positive second derivative of the level density following this gap.]

A fourth model, which is based on the information theory treatment of the random Hamiltonians, is given by Balian (*85*).

10.5 FINITE AND INFINITE MATRICES

Before we end this chapter we mention once again that one respect in which all the matrix models differ from physics is that they all deal with finite matrices, whereas physics deals with infinite matrices. Thus one must always, in interpreting the results of the matrix model, be careful not to use those results which specifically have to deal with the finiteness of the matrix.

Finally, to conclude our discussion, we point out that our review does not include the tremendous volume of work done on the mathematical aspects of the theory of random matrices, and the reader is referred to the literature [see, e.g., Roy (*86*); James (*87*); Krishnaiah and Chang (*88*); Krishnaiah and Waikar (*89*); Krishnaiah and Chattopadhyay (*90*)].

In Appendix A some recent developments on complex multivariate distributions are given.

APPENDIX A
MULTIVARIATE DISTRIBUTIONS

A1. DISTRIBUTIONS

The theory of complex multivariate distributions plays an important role in various areas. As was shown in the text, in nuclear physics these distributions are useful [see, for example, Wigner (*6*), Carmeli (*18*), and Brody *et al.* (*42*)] in studying such problems as the distributions of the spacings between energy levels of nuclei in high excitation. In the area of the multiple time series, these distributions are useful in studying such problems as the structures of the spectral density matrix, since certain suitably defined estimates of the spectral density matrix of the stationary Gaussian multiple time series are approximately distributed as a complex Wishart matrix.

The problems of studying the structure of the above spectral density matrix arise in the analysis of the data in numerous areas. These areas include vibrations of the airframe structures, meteorological forecasts, and signal detection. For some discussion about the usefulness of the complex multivariate distributions in the area of multiple time series, the reader is referred to Hannan (*91*), Liggett (*92, 93*), Priestley, Subba Rao and Tong (*94*), and Brillinger (*95*).

Wooding (*96*), and Goodman (*97*), studied the complex multivariate normal distribution. The joint distributions of the roots of some complex random matrices were derived by James (*98*), Wigner (*99*), and Khatri (*100*), following similar lines as in the analogous real cases. Some work was done in the past, on the distribution problems associated with certain test statistics based on the eigenvalues of the complex Wishart, multivariate beta, multivariate F, and other random matrices. Some of these distribution problems may be solved by following similar lines as in the analogous real cases, whereas some problems need techniques different from the real cases.

In this appendix, following Krishnaiah (*101*), a review of the developments of the complex multivariate distributions is given. We first discuss the evaluation of certain integrals which are needed for the computation of the probability integrals of some complex multivariate distributions. We then review the literature on the distributions of some complex random matrices, as well as the joint densities of the eigenvalues of these random matrices. These random matrices include complex Wishart, multivariate beta, multivariate F, and Gaussian matrices.

The marginal distributions of few roots are subsequently discussed, and the distributions of the traces of some complex random matrices are examined. The distributions of various ratios of the roots of complex Wishart, and multivariate beta matrices, are reviewed. The results on the distributions of the likelihood ratio statistics, for testing the hypotheses on the covariance structures and mean vectors of complex multivariate normal populations, are discussed.

A2. PRELIMINARIES

We start the appendix by evaluating some integrals that are needed in the sequel [Krishnaiah (*101*)].

Let

$$\phi(x_1, \ldots, x_p) = |(y_{ij})|,$$

and

$$\psi(x_1, \ldots, x_p) = |(z_{ij})|,$$

where $y_{ij} = \phi_i(x_j)$ and $z_{ij} = \psi_i(x_j)$. Also, let $\eta(x_1, \ldots, x_p)$ be a symmetric function of the variables $x_1, \ldots, x_p$. Then, one obtains the following formula:

$$\int_{a \leqslant x_1 \leqslant \ldots \leqslant x_p \leqslant b} \cdots \int \eta(x_1, \ldots, x_p)\, \phi(x_1, \ldots, x_p)\, \psi(x_1, \ldots, x_p)\, dx_1 \ldots dx_p = \int_a^b \cdots \int_a^b \eta(x_1, \ldots, x_p)\, |(a_{ij})|\, dx_1 \ldots dx_p, \tag{A2.1}$$

where $a_{ij} = \phi_i(x_j)\, \psi_j(x_j)$.

From Eq. (A2.1) we then obtain the following:

Lemma A2.1 *Let the symmetric function $\eta(x_1, \ldots, x_p)$ be of the form*

$$\eta(x_1, \ldots, x_p) = \sum_m c(m_1, \ldots, m_p)\, x_1^{m_1} \ldots x_p^{m_p}, \tag{A2.2}$$

where $c(m_1, \ldots, m_p)$ is a constant depending upon $m_1, \ldots, m_p$, and the summation is over the values of $m_1, \ldots, m_p$. Then

$$\int_{a \leqslant x_1 \leqslant \ldots \leqslant x_p \leqslant b} \cdots \int \eta(x_1, \ldots, x_p)\, \phi(x_1, \ldots, x_p)\, \psi(x_1, \ldots, x_p)\, dx_1 \ldots dx_p \tag{A2.3}$$

$$= \sum_m c(m_1, \ldots, m_p)\, |B(m_1, \ldots, m_p)|,$$

where

$$B(m_1, \ldots, m_p) = (b_{ij}),$$

and

$$b_{ij} = \int_{a<x<b} x^{m_j} \phi_i(x) \psi_j(x)\, dx.$$

For $\eta(x_1, \ldots, x_p) = 1$, the above lemma was proved by Andreief (*102*).

Next, let $\eta(x_1, \ldots, x_p)$ be any symmetric function of $x_1, \ldots, x_p$. Then, it is seen that

$$\int_{D_1} \cdots \int \eta(x_1, \ldots, x_p)\, \phi(x_1, \ldots, x_p)\, \psi(x_1, \ldots, x_p)\, dx_1 \ldots dx_p$$

$$= \frac{1}{r!\,(p-r)!} \Sigma_1 \Sigma_2 (-1)^{\Sigma\delta_i + \Sigma\alpha_i} \int_{D_2} \cdots \int \eta(x_1, \ldots, x_p)$$

$$\times\ |B_1|\ |B_2|\ dx_1 \ldots dx_p, \qquad \text{(A2.4)}$$

where the domains D_1, D_2 of integration are given by $D_1 : a \leqslant x_1 \leqslant \ldots \leqslant x_r \leqslant x \leqslant x_{r+1} \leqslant \ldots < x_p < b$ and $D_2: a \leqslant x_i \leqslant x (i = 1, \ldots, r)$, $x \leqslant x_j \leqslant b (j = r+1, \ldots, p)$. In Eq. (A2.4), $\delta_1 < \ldots < \delta_r$ is a subset of the integers $1, 2, \ldots, p$ and $\nu_1 < \ldots < \nu_{p-r}$ is the subset complementary to $\delta_1, \ldots, \delta_r$ and Σ_1 denotes the summation over all $\binom{p}{r}$ possible choices of $\delta_1 < \ldots < \delta_r$. Similarly, $\alpha_1 < \ldots < \alpha_r$ is a subset of the integers $1, 2, \ldots, p$ and $\beta_1 < \ldots < \beta_{p-r}$ is the subset complementary to $\alpha_1, \ldots, \alpha_r$, and Σ_2 denotes the summation over $\alpha_1 < \ldots < \alpha_r$. In addition $B_1 = (b_{1gh})$ and $B_2 = (b_{2gh})$, where

$$b_{1gh} = \sum_i \phi_{\delta_g}(x_i)\, \psi_{\alpha_h}(x_i),$$

$$b_{2gh} = \sum_j \phi_{\nu_g}(x_j)\, \psi_{\beta_h}(x_j).$$

But the integral on the right-hand side of Eq. (A2.4) gives:

$$\int_{D_2}\!\!\cdots\!\int \eta(x_1,\ldots,x_p)\,|B_1|\,|B_2|\,dx_1\ldots dx_p$$
$$= \int_{D_2}\!\!\cdots\!\int \eta(x_1,\ldots,x_p)\,|B_1^*|\,|B_2^*|\,dx_1\ldots dx_p\,,$$

where $B_1^* = (b_{1gh}^*)$, $B_2^* = (b_{2gh}^*)$,

$$b_{1gh}^* = \phi_{\delta_g}(x_h)\,\psi_{\alpha_h}(x_h),$$

and

$$b_{2gh}^* = \phi_{\nu_g}(x_{r+h})\,\psi_{\beta_h}(x_{r+h}).$$

Thus, we have the following:

Lemma A2.2 *Let*

$$\eta(x_1,\ldots,x_p) = \Sigma_m\, c(m_1,\ldots,m_p)\, x_1^{m_1}\ldots x_p^{m_p}$$

be a symmetric function of $x_1,\ldots,x_p$. *Then*

$$\int_{D_1}\!\!\cdots\!\int \eta(x_1,\ldots,x_p)\,\phi(x_1,\ldots,x_p)\,\psi(x_1,\ldots,x_p)\,dx_1\ldots dx_p$$

$$= \Sigma_1\,\Sigma_2\,\sum_m\,(-1)^{\Sigma\delta_i+\Sigma\alpha_i}\,c(m_1,\ldots,m_p)\,|B_3|\,|B_4|, \quad \text{(A2.5)}$$

where $B_3 = (b_{3gh})$, $B_4 = (b_{4gh})$, and

$$b_{3gh} = \int_a^x y^{m_h}\,\phi_{\delta_g}(y)\,\psi_{\alpha_h}(y)\,dy,$$

$$b_{4gh} = \int_x^b y^{m_{r+h}}\,\phi_{\nu_g}(y)\,\psi_{\beta_h}(y)\,dy.$$

When $\eta(x_1, \ldots, x_p) = 1$, an alternative expression was given by Khatri (*103*). But, the expression given on the right-hand side of Eq. (A2.5) has some advantages over that given by Khatri from the computational point of view.

Next, let us expand $\phi(x_1, \ldots, x_p)$ as follows:

$$\begin{aligned} \phi(x_1, \ldots, x_p) = {} & \Sigma_3 \Sigma_4 (-1)^{d(r,s)} V(x_1, \ldots, x_r; a_1, \ldots, a_r) \\ & \times V(x_{r+1}, \ldots, x_{r+s}; \alpha_1, \ldots, \alpha_s) \\ & \times V(x_{r+s+1}, \ldots, x_p; \beta_{r+s+1}, \ldots, \beta_p), \end{aligned} \tag{A2.6}$$

where

$$d(r, s) = (r(r+1)/2) + (s(s+1)/2) + \Sigma a_i + \Sigma \alpha_i$$

and

$$V(x_1, \ldots, x_n; b_1, \ldots, b_n) = \begin{vmatrix} \phi_{b_1}(x_1) \ldots \phi_{b_1}(x_n) \\ \vdots \\ \phi_{b_n}(x_1) \ldots \phi_{b_n}(x_n) \end{vmatrix} .$$

Here, $(a_1 < \ldots < a_r)$ is a subset of $(1, \ldots, p)$, and $(t_1, \ldots, t_{p-r})$ is its complementary set. Similarly, $(\alpha_1 < \ldots < \alpha_s)$ is a subset of $(t_1, \ldots, t_{p-r})$ and $(\beta_{r+s+1}, \ldots, \beta_p)$ is its complementary set. In addition, Σ_3 denotes the summation over all possible $\binom{p}{r}$ choices of $(a_1, \ldots, a_r)$, and Σ_4 denotes the summation over all $\binom{p-r}{s}$ possible choices of $(\alpha_1 < \ldots < \alpha_s)$. Similarly, we can express $\psi(x_1, \ldots, x_p)$ as follows:

$$\begin{aligned} \psi(x_1, \ldots, x_p) = {} & \Sigma_5 \Sigma_6 (-1)^{d^*(r,s)} V^*(x_1, \ldots, x_r; \\ & a_1^*, \ldots, a_r^*) \times V^*(x_{r+1}, \ldots, x_{r+s}; \\ & \alpha_1^*, \ldots, \alpha_s^*) \times V^*(x_{r+s+1}, \ldots, x_p); \\ & \beta_{r+s+1}^*, \ldots, \beta_p^*), \end{aligned} \tag{A2.7}$$

where

$$d^*(r, s) = (r(r+1)/2) + (s(s+1)/2) + \Sigma\, a_i^* + \Sigma\, \alpha_i^*,$$

and

$$V^*(x_1, \ldots, x_n; b_1, \ldots, b_n) = \begin{vmatrix} \psi_{b_1}(x_1) \ldots \psi_{b_1}(x_n) \\ \vdots \quad \ldots \quad \vdots \\ \psi_{b_n}(x_1) \ldots \psi_{b_n}(x_n) \end{vmatrix}.$$

Also, $(a_1^* < \ldots < a_r^*)$ is a subset of $(1, \ldots, p)$, and $(t_1^*, \ldots, t_{p-r}^*)$ is its complementary set. Similarly, $(\alpha_1^* < \ldots < \alpha_s^*)$ is a subset of $(t_1^*, \ldots, t_{p-r}^*)$, and $(\beta_{r+s+1}^*, \ldots, \beta_p^*)$ is its complementary set. In addition, Σ_5 denotes the summation over all possible $\binom{p}{r}$ choices of $(a_1^*, \ldots, a_r^*)$ and Σ_6 denotes the summation over all $\binom{p-r}{s}$ possible choices of $(\alpha_1^* < \ldots < \alpha_s^*)$. Using Eqs. (A2.6) and (A2.7) and Lemma A2.1, we then obtain the following:

Lemma A2.3 *Let* $D_3: a \leqslant x_1 \leqslant \ldots \leqslant x_r \leqslant x \leqslant x_{r+1} \leqslant \ldots \leqslant x_{r+s} \leqslant y \leqslant x_{r+s+1} \leqslant \ldots \leqslant x_p \leqslant b$. *Then*

$$\int \ldots \int_{D_3} \phi(x_1, \ldots, x_p)\, \psi(x_1, \ldots, x_p)\, dx_1 \ldots dx_p$$

$$= \Sigma_3\, \Sigma_4\, \Sigma_5\, \Sigma_6\, (-1)^{\Sigma a_i + \Sigma a_i^* + \Sigma \alpha_i + \Sigma \alpha_i^*}\, |B_5|\, |B_6|\, |B_7|, \tag{A2.8}$$

where $B_5 = (b_{5gh})$, $B_6 = (b_{6gh})$, $B_7 = (b_{7gh})$, *and*

$$b_{5gh} = \int_a^x \phi_{a_g}(z)\, \psi_{a_h}{}^*(z)\, dz\,; (g, h = 1, 2, \ldots, r),$$

$$b_{6gh} = \int_x^y \phi_{\alpha_g}(z)\, \psi_{\alpha_h}{}^*(z)\, dz\,; (g, h = r+1, \ldots, r+s),$$

$$b_{7gh} = \int_y^b \phi_{\beta_b}(z)\, \psi_{\beta_h}(z)\, dz\,; (g, h = r+s+1, \ldots, p).$$

Lemma A2.4 *Let* $D_4 : a \leqslant x_1 \leqslant \ldots \leqslant x_r \leqslant x_{r+1} \leqslant x_{r+s} \leqslant x_{r+s+1} \leqslant \ldots \leqslant x_p \leqslant b$. *Then*

$$\int_{D_4} \cdots \int \phi(x_1, \ldots, x_p)\, \psi(x_1, \ldots, x_p)\, dx_1 \ldots dx_r\, dx_{r+s+1} \ldots dx_p$$

$$= \Sigma_3\, \Sigma_4\, \Sigma_5\, \Sigma_6\, (-1)^{\Sigma a_i + \Sigma \alpha_i + \Sigma a_i^* + \Sigma \alpha_i^*}\, V(x_{r+1}, \ldots, x_{r+s};$$

$$\alpha_1, \ldots, \alpha_s) \times V^*(x_{r+1}, \ldots, x_{r+s}; \alpha_1^*, \ldots, \alpha_s^*)$$

$$|B_8|\,|B_9|, \tag{A2.9}$$

where $B_8 = (b_{8gh})$, $B_9 = (b_{9gh})$, *and*

$$b_{8gh} = \int_a^{x_{r+1}} \phi_{a_g}(z)\, \psi_{a_h^*}(z)\, dz\,; (g, h = 1, 2, \ldots, r), \tag{A2.10}$$

$$b_{9gh} = \int_{x_{r+s}}^{b} \phi_{\beta_g}(z)\, \psi_{\beta_h}(z)\, dz\,; (g, h = r+s+1, \ldots, p). \tag{A2.11}$$

A.3 DISTRIBUTIONS OF SOME RANDOM MATRICES

The following definitions will be needed [see James (*98*)] in the sequel [Krishnaiah (*101*)].

Let $\kappa = (k_1, \ldots, k_p)$ be a partition of k such that $k_p \geqslant \ldots \geqslant k_1 \geqslant 0$ and $k_1 + \ldots + k_p = k$. Also, let $\tilde{C}_\kappa(A)$ denote the zonal polynomial of a Hermitian matrix A of order $p \times p$. The hypergeometric functions with matrix arguments are as given below:

$${}_r\tilde{F}_q(c_1, \ldots, c_r; d_1, \ldots, d_q; A)$$

$$= \sum_{k=0}^{\infty} \sum_{\kappa} \frac{(c_1)_\kappa \ldots (c_r)_\kappa\, \tilde{C}_\kappa(A)}{(d_1)_\kappa \ldots (d_q)_\kappa\, k!}, \tag{A3.1}$$

$${}_r\tilde{F}_q(c_1, \ldots, c_r; d_1, \ldots, d_q; G, H)$$

$$= \sum_{k=0}^{\infty} \sum_{\kappa} \frac{(c_1)_\kappa \ldots (c_r)_\kappa \tilde{C}_\kappa(G) \tilde{C}_\kappa(H)}{(d_1)_\kappa \ldots (d_q)_\kappa \tilde{C}_\kappa(I_p)\, k!}, \qquad \text{(A3.2)}$$

where $c_1, \ldots, c_r$, and $d_1, \ldots, d_q$ are real or complex constants. Here, we note that ${}_0\tilde{F}_0(A) = \text{eTr}\, A$ and ${}_1\tilde{F}_0(a; A) = |I - A|^{-a}$. When $A = 0$, the right-hand side of Eq. (A3.1) is equal to 1. Also, the right-hand side of Eq. (A3.2) is equal to 1 when $G = 0$ or $H = 0$. Throughout this appendix, eTr B denotes the exponential of the trace of B, and

$$(a)_\kappa = \prod_{i=1}^{p} (a - i + 1)_{\kappa},$$

$$(a)_\kappa = a(a+1) \ldots (a + \kappa - 1),$$

$$\tilde{\Gamma}_p(a) = \pi^{p(p-1)/2} \prod_{i=1}^{p} (a - i + 1)\, \kappa_i.$$

The zonal polynomials of the matrix variables were first considered by Hua (*104*) and later by James (*105*) independently. The general system of hypergeometric functions with matrix arguments is due to Herz (*106*) who defined them in integral forms. Since the computation of the expressions involving zonal polynomials is complicated, it would be of interest to try to get approximations to the above hypergeometric functions, starting with the expressions of Herz.

We now review the results on the joint distributions of the roots of some random matrices.

Let $Z = X + iY$, where X and Y are random matrices of order $m \times p$. Also, let the rows of (X^t, Y^t) be distributed independently as $2p$-variate normal with mean vector $(\mu_1{}^t, \mu_2{}^t)$ and covariance matrix E, where E is given by

$$E = \begin{bmatrix} E_1 & E_2 \\ -E_2 & E_1 \end{bmatrix}. \qquad \text{(A3.3)}$$

Then, the rows of Z are known to be distributed as complex multivariate normal.

Now, let $A_1 = ZZ^\dagger$, where $Z^\dagger$ is the transpose of the complex conjugate of Z, $Z^\dagger = \bar{Z}^t$. Then, the distribution of A_1 is known to be a central (noncentral) complex Wishart matrix with m degrees of freedom when $M = 0$ ($M \neq 0$), where $E(Z) = M$. Also, $E(A_1/m) = \Sigma_1 = 2(E_1 - iE_2)$. Next, let $A_2 : p \times p$ be a central complex Wishart matrix with n degrees of freedom, and $E(A_2/n) = \Sigma_2$. Then $F = A_1 A_2^{-1}$ is known to be a central (noncentral) complex multivariate F matrix when $M = 0$ ($M \neq 0$). Similarly, $B = A_1(A_1 + A_2)^{-1}$ is said to be a central or noncentral complex multivariate beta matrix according to whether $M = 0$ or $M \neq 0$.

Next, let $E_2 = 0$, $E_1 = \text{diag.}\ (\lambda_1, \ldots, \lambda_p)$,

$$B_0 = ((A - 2mE_1)(2m)^{1/2}) = B_1 + iB_2,$$

where $B_1 = (b_{1ij})$ and $B_2 = (b_{2ij})$. When $m \to \infty$, the random variables b_{1ij} and b_{2ij} by central limit theorem, are distributed independently and normally with zero means and variances given by $E(b_{1ij}^2) = \lambda_i \lambda_j (i \neq j)$, $E(b_{1ij}^2) = 2\lambda_i^2$, and $E(b_{2ij}^2) = \lambda_i \lambda_j$; here, we note that $b_{2ii} = 0$.

Now, let $A = (a_{ij}) = R + iS$, where $A : p \times p$ is a Hermitian random matrix, $R = (r_{ij})$ and $S = (s_{ij})$. Then $s_{ii} = 0$. Now, let the elements of R and the off-diagonal elements of S be distributed independently. We assume that the variances of the off-diagonal elements of R and S are equal to 1, and the variances of the diagonal elements of R are equal to 2. Then, we refer to $A = R + iS$ as the central or noncentral complex Gaussian matrix according to whether $E(A) = 0$ or $E(A) \neq 0$.

The complex multivariate normal distribution was derived by Wooding (*96*). The density function of the complex multivariate normal is given by

$$f(Z) = \frac{\text{eTr}[-\Sigma^{-1}(Z - M)(Z - M)^\dagger]}{\pi^{pm}\ |\Sigma|^m}. \tag{A3.4}$$

The distribution of the complex Wishart matrix is known [see Goodman (*97*)] to be given by

$$f(A_1) = \frac{|A_1|^{m-p}}{\tilde{\Gamma}_p(m)\,|\Sigma|^m}\ \mathrm{eTr}(-\tfrac{1}{2}\ \Sigma_1^{-1} A_1), \tag{A3.5}$$

where $M = 0$. When $M \neq 0$, the distribution of A_1 is known to be given by

$$f(A_1) = \mathrm{eTr}(-\Sigma^{-1} MM^\dagger)\ {}_0\tilde{F}_1\,(m\,;\Sigma^{-1} MM^\dagger\,\Sigma^{-1} A_1)$$

$$\times\ \frac{|A_1|^{m-p}}{\tilde{\Gamma}_p(m)\,|\Sigma|^m}\ \mathrm{eTr}(-\Sigma^{-1} A_1). \tag{A3.6}$$

Srivastava (*107*) gave a simplified derivation of the central complex Wishart distribution.

Let $w_p \geqslant \ldots \geqslant w_1$ be the latent roots of the complex Wishart matrix $A_1\,\Sigma^{-1}$ with m degrees of freedom. The joint density of the roots $w_1, \ldots, w_p$ is given by

$$h_1(w_1, \ldots, w_p) = \mathrm{eTr}(-\Omega)\ {}_0\tilde{F}_1\,(m\,;\Omega\,;A_1)$$

$$\times\ \frac{\pi^{p(p-1)}}{\tilde{\Gamma}_p(m)\,\tilde{\Gamma}_p(p)}\ \mathrm{eTr}(-A_1)\,|A_1|^{m-p}$$

$$\prod_{i>j}^{p}\ (w_i - w_j)^2, \tag{A3.7}$$

where Ω is a diagonal matrix whose elements are the roots of $MM^\dagger\,\Sigma^{-1}$. When $M = 0$, the joint density of the roots $(a_1 \leqslant \ldots \leqslant a_p)$ of A_1 is given by

$$h_1(a_1, \ldots, a_p) = \frac{\pi^{p(p-1)}}{\tilde{\Gamma}_p(m)\,\tilde{\Gamma}_p(p)}\ \frac{\mathrm{eTr}(-\Sigma^{-1} A_1)}{|\Sigma|^m}$$

$$\times\ |A_1|^{m-p}\ \prod_{i>j}^{p}\ (a_i - a_j)^2. \tag{A3.8}$$

Next, let us assume that the rank of A_1 is q and let the nonzero roots of $A_1 A_2^{-1}$ be $f_q \geqslant \ldots \geqslant f_1$. When $m \leqslant p$, and $\Sigma = \Sigma_2$ the joint density of $f_1, \ldots, f_q$ is given by

$$h_2(f_1, \ldots, f_q) = \text{eTr}(-\Omega)\ {}_1\tilde{F}_1\,(m+n;\ p;\ \Omega, (I+F^{-1})^{-1})$$

$$\times \frac{\pi^{m(m-1)}\,\tilde{\Gamma}_m(m+n)}{\tilde{\Gamma}_m(p)\,\tilde{\Gamma}_m(m+n-p)\,\tilde{\Gamma}_m(m)}$$

$$\frac{|F|^{p-m}}{|I+F|^{m+n}} \prod_{i>h} (f_i - f_j)^2, \tag{A3.9}$$

where $F = A_1 A_2^{-1}$ and $\Omega = M^\dagger \Sigma^{-1} M$. When $m \geqslant p$, the joint density of $f_1, \ldots, f_q$ is given by Eq. (109) in James (*98*).

Now, let $a_p \geqslant \ldots \geqslant a_1$ be the latent roots of the Gaussian matrix A. Then, the joint density of $a_1, \ldots, a_p$ is known [see Waikar, Chang, and Krishnaiah (*108*)] to be given by

$$h_3(a_1, \ldots, a_p) = C\,\text{eTr}(-\frac{1}{2} M^2)$$

$$\times \sum_{k=0}^{\infty} \sum_{\kappa} \frac{\tilde{C}_\kappa\left(\frac{1}{2}M\right)}{k!\,\tilde{C}_\kappa(I_p)}\,\tilde{C}_\kappa(A) \prod_{i=1}^{p} e^{-a_i^2/4}$$

$$\times \prod_{i>j=1}^{p} (a_i - a_j)^2;\quad (-\infty \leqslant A_i \leqslant \infty). \tag{A3.10}$$

Also, let

$$C = \pi^{p(p-1)}/\tilde{\Gamma}_p(p)\, 2^{(p^2+p)/2}\, \pi^{p^2/2}.$$

When $M = 0$, the joint density of $a_1, \ldots, a_p$ is known [see Wigner (*99*)] to be given by

$$h_3(a_1, \ldots, a_p) = C \prod_{i=1}^{p} \text{e}^{-a_i^2/4} \prod_{i>j=1}^{p} (a_i - a_j)^2. \tag{A3.11}$$

Next, let the m rows of $(Z_1 \vdots Z_2)$ be distributed independently as $(p+q)$ variate complex normal with zero means and covariance matrix

$$\begin{bmatrix} \Sigma_{11} & \Sigma_{12} \\ \Sigma_{21} & \Sigma_{22} \end{bmatrix},$$

and let $p \leqslant q$. Also, let $r_p^2 \geqslant \ldots \geqslant r_1^2$ be the latent roots of $R^2 = (Z_1 Z_1^\dagger)^{-1} (Z_1 Z_2^\dagger) (Z_2 Z_2^\dagger)^{-1} Z_2 Z_1^\dagger$. Then the joint density of $r_1^2, \ldots, r_p^2$ is known to be given by

$$h_4(r_1^2, \ldots, r_p^2) = |I - P^2|^m \, {}_2\tilde{F}_1(m, m; \; q; \; P^2, R^2)$$
$$\times \frac{\tilde{\Gamma}_p(m) \, \pi^{p(p-1)}}{\tilde{\Gamma}_p(m-q) \, \tilde{\Gamma}_p(q) \, \tilde{\Gamma}_p(p)}$$
$$|R^2|^{q-p} \, |I - R^2|^{m-q-p}$$
$$\times \prod_{i>j} (r_i^2 - r_j^2)^2, \qquad \text{(A3.12)}$$

where $P^2 = \Sigma_{11}^{-1} \Sigma_{12} \Sigma_{22}^{-1} \Sigma_{21}$. The above formula was given in James (*98*).

When $M = 0$, we will refer to the distribution of A_1 as the central complex Wishart matrix or the central complex Wishart matrix with $\Sigma_1 \neq I$, accordingly as $\Sigma = I$ or $\Sigma \neq I$. Similarly, when $M = 0$, we will refer to the distribution of $A_1 (A_1 + A_2)^{-1}$ as the central multivariate beta matrix (central multivariate beta matrix with $\Sigma \neq \Sigma_2$) when $\Sigma = \Sigma_2 (\Sigma \neq \Sigma_2)$. Goodman (*97*) expressed the densities of multiple coherence and partial coherence as infinite series, while Kabe (*109*) expressed them as finite series.

Now, let Z^t be partitioned as $Z^t = (Z_1{}^t, \ldots, Z_q{}^t)$, where Z_j is of order $m \times p_j$ and the distribution of Z is given by Eq. (A3.4). In addition, let $S_j = Z_j Z_j^\dagger A_j$, where the elements of A_j's are constants. Then, the joint characteristic function of $S_1, \ldots, S_q$ is given by

$$\phi(\theta_1, \ldots, \theta_q) = E \left\{ \mathrm{eTr}\, i(\theta_1 S_1 + \ldots + \theta_q S_q) \right\}$$
$$= |I - A^* \Sigma|^{-m} \, \mathrm{eTr} \left\{ A^* (I - \Sigma A^*)^{-1} M M^\dagger \right\}, \qquad \text{(A3.13)}$$

where $A^* = \text{diag.}\,(A_1{}^*, \ldots, A_q{}^*)$, and $A_j{}^* = i\theta_j A_j$. If $p_1 = \ldots = p_q$, the characteristic function of $S_1 + \ldots + S_q$ is obtained from the above equation by choosing $\theta_1, \ldots, \theta_q$ to be equal. The joint characteristic function of $y_1, \ldots, y_q$, where $y_j = \text{Tr}\, S_j (j = 1, \ldots, q)$, is given by

$$|I - B^*\Sigma|^{-m} \text{ eTr} \left\{B^* (I - \Sigma B^*)^{-1} MM^\dagger \right\}, \tag{A3.14}$$

where $B^* = \text{diag.}\,(B_1{}^*, \ldots, B_q{}^*)$, and $B_j{}^* = \text{it}_j A_j$. The case $m = 1$ and $q = 1$ of Eq. (A3.14) is given by Turin (*110*).

A4. MARGINAL DISTRIBUTIONS OF FEW ROOTS

Let $l_1, \ldots, l_p$ be the latent roots of a class of random matrices, and let the joint density of these roots be of the form

$$h(l_1, \ldots, l_p) = C \prod_{i=1}^{p} \psi(l_i) \sum_{k=0}^{\infty} \sum_{\kappa} c(\kappa, \Omega)\, |(l_j^{i-1})|$$

$$\times\, |(l_j^{i-1+k_{p-i+1}})|;$$

$$(a \leqslant l_1 \leqslant \ldots \leqslant l_p \leqslant b), \tag{A4.1}$$

where C is a constant, $\kappa = (k_1, \ldots, k_p)$ is a partition of k such that $k_p \geqslant \ldots \geqslant k_1$, $\psi(x)$ is a function of x, and $c(\kappa, \Omega)$ depends on κ and the population parameter matrix Ω. The joint densities of the eigenvalues of the random matrices considered in the preceding section are special cases of Eq. (A4.1) [Krishnaiah (*101*)].

The probability integral of the joint distribution of the extreme roots l_1 and l_p is given by

$$P[c \leqslant l_1 \leqslant l_p \leqslant d] = C \sum_{k=0}^{\infty} \sum_{\kappa} c(\kappa, \Omega)\, |A|, \tag{A4.2}$$

where $A = (a_{ij})$ and

$$a_{ij} = \int_{c \leqslant x \leqslant d} \psi(y)\, y^{i+j+k_{p-i+1}-2}\, dy.$$

Equation (A4.2) follows immediately by applying Lemma A2.1.

The c.d.f. of the largest root l_p is obtained by putting $c = a$ in Eq. (A4.2). The c.d.f. of the smallest root l_1 is given by

$$P[l_1 \leqslant c] = 1 - P[c \leqslant l_1 \leqslant l_p \leqslant b], \tag{A4.3}$$

where the right side of Eq. (A4.3) can be evaluated by applying Eq. (A4.2). The c.d.f. of the intermediate root l_s $(2 \leqslant s \leqslant p-1)$ is given by

$$P[l_s \leqslant c] = P[l_{s+1} \leqslant c] + P[a \leqslant l_1 \leqslant \ldots \leqslant l_s \leqslant c \leqslant l_{s+1} \leqslant \ldots \leqslant l_p \leqslant b], \tag{A4.4}$$

where

$$a \leqslant l_1 \leqslant \ldots \leqslant l_p \leqslant b.$$

But, applying Lemma A2.2, we obtain

$$P[l_1 \leqslant \ldots \leqslant l_s \leqslant x \leqslant l_{s+1} \leqslant \ldots \leqslant l_p]$$

$$= C \sum_{k=0}^{\infty} \sum_{\kappa} \Sigma_1 \Sigma_2 (-1)^{\Sigma\alpha_i + \Sigma\delta_i} |B_3|\,|B_4|, \tag{A4.5}$$

where Σ_1, Σ_2, α_i's and δ_i's were defined by Eq. (A2.4), $B_3 = (b_{3gh})$ and $B_4 = (b_{4gh})$. Here

$$b_{3gh} = \int_a^x \phi_{\delta_g}(y)\, \psi_{\alpha_h}(y)\, dy,$$

$$b_{4gh} = \int_x^b \phi_{\nu_g}(y)\, \psi_{\beta_h}(y)\, dy,$$

and

$$\phi_i(y) = y^{i-1},$$

$$\psi_i(y) = \psi(y) y^{k_{p-i+1}+i-1}.$$

The joint density of $l_{r+1}, \ldots, l_{r+s}$ follows immediately by applying Lemma A2.4.

Similarly, one can derive the joint density of any few ordered roots that are not necessarily consecutive, but the resulting expressions are complicated. We will now discuss the joint probability integral associated with any pair of roots l_r, l_s $(1 \leqslant r \leqslant s \leqslant p)$.

We know that

$$\begin{aligned} P[x_1 \leqslant l_r \leqslant l_s \leqslant x_2] = \; & P[l_{r-1} \leqslant x_1 \leqslant l_r \leqslant l_s \leqslant x_2 \leqslant l_{s+1}] \\ & + P[x_1 \leqslant l_{r-1} \leqslant l_r \leqslant l_s \leqslant x_2 \leqslant l_{s+1}] \\ & + P[l_{r-1} \leqslant x_1 \leqslant l_r \leqslant l_s \leqslant l_{s+1} \leqslant x_2] \\ & + P[x_1 \leqslant l_{r-1} \leqslant l_r \leqslant l_s \leqslant l_{s+1} \leqslant x_2]. \end{aligned} \tag{A4.6}$$

Each quantity on the right-hand side of Eq. (A4.6) can be evaluated using Lemma A2.3.

Khatri (*111*) gave expressions for the extreme roots of the central complex Wishart and multivariate beta matrices, whereas Al-Ani (*112*) derived the expressions for the intermediate roots of the above random matrices. Also, Khatri (*113*) derived the distributions of the individual roots of the central Wishart matrix A_1 with $\Sigma \neq I$, noncentral multivariate beta matrix, central multivariate beta matrix with $\Sigma \neq \Sigma_2$, and noncentral canonical correlation matrix in the complex cases. The method used in the above papers for obtaining the distributions is different from the method used in the present appendix. The expressions given in this appendix for the distributions of the intermediate roots are better, from the computational point of view, than the corresponding expressions given in Khatri (*111, 113*) and Al-Ani (*112*).

Approximate percentage points of the largest root of the Wishart matrix and multivariate beta matrix were given by Pillai and Young (*114*) and Pillai and Jouris (*115*), respectively, in the central complex cases. Exact percentage points of the smallest root and intermediate roots of the central complex Wishart matrix were constructed by Schuurmann and Waikar (*116*), and Krishnaiah and Schuurmann (*117*). Krishnaiah and Schuurmann (*117*) also constructed the exact percentage points of the distributions of the individual roots of the central complex multivariate beta matrix. Exact percentage points of the joint distribution of the extreme roots of the central complex Wishart matrix as well as that of the central complex multivariate beta matrix were given by Krishnaiah and Schuurmann (*118*).

The joint densities of any few unordered roots of the complex Wishart, complex multivariate beta matrix, and complex Gaussian matrix, were known [see Wigner (*99*), and Mehta (*119*)] in the literature for the central cases. Waikar, Chang, and Krishnaiah (*108*) derived the joint densities of any few unordered roots of the complex Gaussian ensemble matrix, complex Wishart matrix A_1 with $\Sigma \neq I$, complex multivariate beta matrix, and complex canonical correlation matrix in the noncentral cases. We now discuss the joint distribution of any few unordered roots of a class of random matrices.

Let $l_1, \ldots, l_p$ be the unordered roots of a class of random matrices and let their joint density be given by

$$f_1(l_1, \ldots, l_p) = (C/p!) \prod_{i=1}^{p} \psi(l_i) \sum_{k=0}^{\infty} \sum_{\kappa} c(\kappa, \Omega) \times |(l_j^{i-1})| \, |(l_j^{i-1+k_{p-i+1}})| ;$$

$$(a \leqslant l_i \leqslant b), \; (i = 1, \ldots, p). \tag{A4.7}$$

Then, the joint density of $l_1, \ldots, l_r$ is given by

$$f_2(l_1, \ldots, l_r) = \int_a^b \cdots \int_a^b f(l_l, \ldots, l_p) \, dl_{r+1} \ldots dl_p . \tag{A4.8}$$

The right-hand side of Eq. (A4.8) can be evaluated using the following lemma:

Lemma A4.1 *Let* $\phi(x_1, \ldots, x_p)$ *and* $\psi(x_1, \ldots, x_p)$ *be as defined in Section A.2. Then*

$$\int_a^b \cdots \int_a^b \phi(x_1, \ldots, x_p) \, \psi(x_1, \ldots, x_p) \, dx_{r+1} \ldots dx_p = \Sigma_1 \, \Sigma_2 \, (-1)^{\Sigma \delta_i + \Sigma \alpha_i} |B_1| \, |B_5| , \tag{A4.9}$$

where B_1, δ_i's, α_i's and the summations Σ_1 and Σ_2 are as defined in Eq. (A2.4), and $B_5 = (b_{5gh})$, where

$$b_{5gh} = (p-r)! \int_a^b \phi_{\nu_g}(y)\, \psi_{\beta_h}(y)\, dy. \tag{A4.10}$$

A5. MOMENTS OF THE ELEMENTARY SYMMETRIC FUNCTIONS

Let the joint density of the roots $l_1 < \ldots < l_p$ of a random matrix be given by Eq. (A4.1). Also, let [Krishnaiah (*101*)]

$$\zeta_q(l_1, \ldots, l_p) = \sum_i l_{i_1} \ldots l_{i_q}; \quad (q \leqslant p),$$

where the summation is over all possible values of $i_1 < \ldots < i_q$. Then,

$$\{\zeta_q(l_1, \ldots, l_p)\}^s = \sum_s \frac{s!}{s_1! \ldots s_{p^*}!} l_1^{\eta_1} \ldots l_p^{\eta_p},$$

where $p^* = \binom{p}{q}$, $s = \Sigma_{i=1}^{p^*} s_i$, and $\eta_1, \ldots, \eta_p$ depend upon $s_1, \ldots, s_{p^*}$; the summation is over all possible values of $s_1, \ldots, s_{p^*}$. Then, the sth moment of $\zeta_q(l_1, \ldots, l_p)$ is given by

$$E\{\zeta_q(l_1, \ldots, l_p)\}^s = \int_{a \leqslant l_1 \leqslant \ldots \leqslant l_p \leqslant b} \cdots \int \{\zeta_q(l_1, \ldots, l_p)\}^s f(l_1, \ldots, l_p)\, dl_1 \ldots dl_p,$$

where $f(l_1, \ldots, l_p)$ is given by Eq. (A4.1). Applying now Lemma A2.1, we then obtain the following result:

Lemma A5.1 *The* sth *moment of* qth *order elementary symmetric function* $\zeta_q(l_1, \ldots, l_p)$ *is given by*

$$E\{\zeta_q(l_1, \ldots, l_p)\}^s = C \sum_k \sum_\kappa \sum_s c(\kappa, \Omega) \frac{s!}{s_1! \ldots s_{p^*}!} |B|, \tag{A5.1}$$

where $B = (b_{ij})$ and b_{ij} is given by

$$b_{ij} = \int_a^b \psi(x) x^{i+j+\eta_j+k_{p-i+1}-2}\, dx. \tag{A5.2}$$

When $a = 0$, $b = 1$, and $\psi(x) = x^n(1-x)^q$, Eq. (A5.2) then reduces to

$$b_{ij} = \beta(n+i+j+\eta_j+k_{p-i+1}-1, q+1).$$

For $a = 0$, $b = \infty$, and $\psi(x) = e^{-x} x^n$, one obtains

$$b_{ij} = \beta(n+i+j+\eta_j+k_{p-i+1}-1, q+1).$$

If $a = -\infty$, $b = \infty$, and $\psi(x) = e^{-x^2/4}$, then

$$b_{ij} = 0, \text{ (if } i+j+\eta_j+k_{p-i+1}-2 \text{ is odd)},$$

and

$$b_{ij} = 4^{(i+j+\eta_j+k_{p-i+1}-1)/2}\, \Gamma[(i+j+\eta_j+k_{p-i+1}-2)/2],$$
$$\text{(if } i+j+\eta_j+k_{p-i+1}-2 \text{ is even)}.$$

The Laplace transformation of the statistic $T = \Sigma_{i=1}^p l_i$ is given by

$$\mathscr{L}(t;T) = \int \ldots \int_{a \leqslant l_1 \leqslant \ldots \leqslant l_p \leqslant b} e^{-tT} f(l_1, \ldots, l_p)\, dl_1 \ldots dl_p$$
$$= C \sum_{k=0}^{\infty} \sum_{\kappa} c(\kappa, \Omega)\, |B^*|, \tag{A5.3}$$

where $B^* = (b_{ij}^*)$ and b_{ij}^* is given by

$$b_{ij}^* = \int_a^b e^{-tx} x^{i+j+k_{p-i+1}-2} \psi(x)\, dx. \tag{A5.4}$$

We now derive the distribution of T when $a = 0$, $b = 1$, and $\psi(x) = x^n (1 - x)^q$. In this special case, one obtains

$$b_{ij}^* = \sum_{l=0}^{\infty} (-1)^l \binom{q}{l} \frac{(l+i+j+n-1+k_{p-i+1})!}{t^{l+i+j+n-1+k_{p-i+1}}} \left[1 - e^{-t} \sum \frac{t^m}{m!}\right], \tag{A5.5}$$

where the summation in the square bracket is over values of m from 0 to $l + i + j + n - 1 + k_{p-i+1}$; here, we note that $\binom{q}{l} = 0$ when q is integer and is less than l. The Laplace transformation is of the form

$$\mathscr{L}(t;T) = C \sum_{k=0}^{\infty} \sum_{\kappa} c(\kappa, \Omega) \left\{ \sum_i d_i \frac{e^{-t\alpha_i}}{t^{\beta_i}} \right\}, \tag{A5.6}$$

where the coefficients d_i, α_i, and β_i depend on n, q, p, and the elements of the partition κ. Now, inverting the right-hand side of Eq. (A5.6), we obtain the following expression for the density of T:

$$g(T) = C \sum_{k=0}^{\infty} \sum_{\kappa} c(\kappa, \Omega) \left\{ \sum_i d_i \frac{(T - \alpha_i)_+^{\beta_i - 1}}{(\beta_i - 1)!} \right\}; \quad (0 \leqslant T \leqslant p), \tag{A5.7}$$

where $(x)_+$ is equal to x or 0 according as $x \geqslant 0$ or $x < 0$.

We will now consider the distribution of $T_1 = \Sigma_i [l_i/(1 - l_i)]$. The Laplace transformation of T_1 is given by

$$\mathscr{L}(t;T_1) = \int_0^{\infty} e^{-tT_1} f(T_1)\, dT_1 = C \sum_{k=0}^{\infty} \sum_{\kappa} c(\kappa, \Omega) \, |D|, \tag{A5.8}$$

where $D = (d_{ij})$ and d_{ij} is given by

$$d_{ij} = \int_a^b e^{-tx/(1-x)} x^{i+j+k_{p-i+1}-2} \psi(x)\, dx.$$

Now, let $a = 0$, $b = 1$, and $\psi(x) = x^n (1 - x)^q$. Then the above formula gives

$$d_{ij} = \int_0^{\infty} e^{-tz} \frac{z^{n+i+j+k_{p-i+1}-2}}{(1+z)^{n+q+i+j+k_{p-i+1}}} dz = \mathscr{L}(t; \eta(i, j; z)), \tag{A5.9}$$

where the function $\eta(i, j; z)$ is given by

$$\eta(i, j; z) = \frac{z^{n+i+j+k_{p-i+1}-2}}{(1+z)^{n+q+i+j+k_{p-i+1}}} . \tag{A5.10}$$

Thus, in this special case, we obtain

$$\mathscr{L}(t; T_1) = C \sum_{k=0}^{\infty} \sum_{\kappa} c(\kappa, \Omega) \left\{ \Sigma \pm \mathscr{L}(t; \eta(i_1, i_2; z)) \ldots \right.$$
$$\left. \mathscr{L}(t; \eta(i_{p-1}, i_p; z)) \right\} , \tag{A5.11}$$

where the summation inside the curly bracket is taken over all the permutations $(i_1, \ldots, i_p)$ of $(1, \ldots, p)$ with a plus sign if $(i_1, \ldots, i_p)$ is an even permutation, and a minus sign if it is an odd permutation. Thus, the distribution of T_1 is given by

$$g_1(T_1) = C \sum_{k=0}^{\infty} \sum_{\kappa} c(\kappa, \Omega) \left\{ \Sigma \pm \eta(i_1, i_2; T_1) * \ldots * \eta(i_{p-1}, \right.$$
$$\left. i_p; T_1) \right\} , \tag{A5.12}$$

where * in Eq. (A5.12) denotes convolution.

The distributions of the traces of the central complex bivariate beta matrix and the central complex bivariate F matrix were derived by Pillai and Jouris (*120*). Using Eq. (A5.7), exact percentage points of the distribution of the trace of the central complex multivariate beta matrix were computed by Krishnaiah and Schuurmann (*121*); these authors (*122*) have also given new tables by approximating the above distribution with Pearson's Type I distribution. The accuracy of this approximation is satisfactory for several practical situations.

Krishnaiah and Schuurmann (*122*) also constructed tables for the distribution of the trace of the central complex multivariate F matrix by approximating this distribution with a suitable Pearson type distribution.

A6. DISTRIBUTIONS OF THE RATIOS OF THE ROOTS

Let the joint density of the roots be given by Eq. (A4.1) and let $a \geqslant 0$. Making the transformations $l_p = l_p$ and $l_i = f_{ip}\, l_p$ for $i = 1, 2, \ldots, p-1$ in Eq. (A4.1), and integrating out l_p, we obtain the following expression for the joint density of $f_{1p}, \ldots, f_{p-1,p}$ [Krishnaiah (*101*)]:

$$g_1(f_{1p}, \ldots, f_{p-1,p}) = C \sum_k \sum_\kappa c(\kappa, \Omega)\, |(f_{jp}^{i-1})| \times |(f_{jp}^{i-1+k_{p-i+1}})| \int_a^b \prod_{i=1}^{p} \psi(l_p f_{ip})\, l_p^{k+p^2-1}\, dl_p . \tag{A6.1}$$

If we make the transformations $f_i = l_i/\Sigma l_j$ for $i = 1, 2, \ldots, p-1$ and $f_p = \Sigma l_i$ in Eq. (A4.1), and integrate out f_p, we obtain the following expression for the joint density of $f_1, \ldots, f_{p-1}$:

$$g_2(f_1, \ldots, f_{p-1}) = C \sum_k \sum_\kappa c(\kappa, \Omega)\, |(f_j^{*i-1})|\, |(f_j^{*i-1+k_{p-i+1}})| \times \int_a^b f_p^{k+p^2-1} \prod_{i=1}^{p-1} \psi(f_i f_p)\, \psi\left[f_p\left(1 - \sum_{i=1}^{p-1} f_i\right)\right] df_p , \tag{A6.2}$$

where f_i^* is given by

$$f_p^* = 1 - \Sigma_{i=1}^{p-1} f_i,$$

and $f_i^* = f_i$ for $i = 1, 2, \ldots, p-1$.

Next, let us make the transformations $f_{i,i+1} = l_{i+1}/l_i$ for $i = 1, 2, \ldots, p-1$ and $l_1 = l_1$ in Eq. (A4.1), and integrate out l_1. Then, we obtain the following expression for the joint density of $f_{12}, \ldots, f_{p-1,p}$:

$$g_3(f_{12}, \ldots, f_{p-1,p}) = C \sum_k \sum_\kappa c(\kappa, \Omega) \prod_{i=1}^{p-2} f_{i,i+1}^{p-i-1}$$

$$\times \left| \left(\prod_{i=0}^{j-1} f_{i,i+1} \right)^{i-1} \right|$$

$$\left| \left(\prod_{i=0}^{j-1} f_{i,i+1} \right)^{i-1+k_{p-i+1}} \right|$$

$$\times \int_a^b l_1^{p-1+p(p-1)k} \prod_{i=1}^{p}$$

$$\psi \left[l_1 \left(\prod_{j=0}^{i-1} f_{j,j+1} \right) \right] dl_1 . \qquad \text{(A6.3)}$$

Similarly, we can obtain the joint distributions of other ratios like $f_{21}, \ldots, f_{p1}$ or $f_2, \ldots, f_p$.

When $l_p \geqslant \ldots \geqslant l_1$ are the roots of the central complex Wishart matrix, Krishnaiah and Schuurmann (*123*) derived the exact marginal distributions of the statistics l_1/l_p and $l_i/\Sigma_{j=1}^p l_j$ for $i = 1, \ldots, p$, and computed the percentage points of these statistics. Krishnaiah and Schuurmann (*124*) also derived the exact distribution of the ratio of the extreme roots of the central complex multivariate beta matrix, and computed percentage points of this distribution.

A7. DISTRIBUTIONS OF THE LIKELIHOOD RATIO TEST STATISTICS

Following the same lines as in the real case, Goodman (*125*) showed that the distribution of the determinant of the central complex Wishart matrix is the product of the distributions of the central chi-square variates.

Giri (*126*) proved some optimum properties of the likelihood ratio tests under the hypothesis that the mean vector is equal to a specified value, and the hypothesis of independence of one variable with a set of variables when the underlying distribution is complex multivariate normal. Wahba (*127*) and Pillai and Nagarsenker (*128*) considered the distribution of the likelihood ratio test statistic for sphericity of the complex multivariate normal population. Gupta (*129*) computed exact percentage points of the distribution of the determinant of the central complex multivariate beta matrix for some special cases. We will now briefly review the recent work of Krishnaiah, Lee, and Chang on the distributions of the likelihood ratio statistics for testing certain hypotheses [Krishnaiah (*101*)].

Let $Z' = (Z_1', \ldots, Z_q')$ be distributed as a complex multivariate normal population with mean vector $\mu' = (\mu_1', \ldots, \mu_q')$ and covariance matrix Σ, and let Z_i be of order $p_i \times 1$. Also, let

$$E\left\{(Z_i - \mu_i)(Z_j - \mu_j)^\dagger\right\} = \Sigma_{ij}.$$

In addition, let H_1, H_2, H_3, and H_4 denote the following hypotheses:

$$H_1 : \Sigma_{ij} = 0; \qquad (i \neq j = 1, \ldots, q),$$

$$H_2 : \Sigma = \sigma^2 \Sigma_0,$$

$$H_3 : \Sigma = \Sigma_0,$$

$$H_4 : \Sigma = \Sigma_0, \qquad \mu = \mu_0,$$

where σ^2 is unknown, and μ_0 and Σ_0 are known.

If we denote the likelihood ratio test statistics for H_1, H_2, H_3, and H_4 by λ_1, λ_2, λ_3, and λ_4, respectively, then it is known that

$$\lambda_1 = \frac{|A|}{\prod_{i=1}^{q} |A_{ii}|}, \tag{A7.1}$$

$$\lambda_2 = \frac{|A\Sigma_0^{-1}|}{(\mathrm{Tr}\, A\Sigma_0^{-1}/s)^s}, \tag{A7.2}$$

$$\lambda_3 = (e/n)^{sn} \, |A\Sigma_0^{-1}|^n \, \mathrm{eTr}(- A\Sigma_0^{-1}), \tag{A7.3}$$

$$\lambda_4 = (e/N)^{sN} \, |A\Sigma_0^{-1}|^N \, \mathrm{eTr}[- \Sigma_0^{-1} \{A + N(Z. - \mu_0)(Z. - \mu_0)^\dagger\}], \tag{A7.4}$$

where $s = \Sigma_{i=1}^{q} p_i$, and $n = N - 1$. In the above equations the matrix A is defined by

$$A = \begin{bmatrix} A_{11} & A_{12} & \cdots & A_{1q} \\ A_{21} & A_{22} & \cdots & A_{2q} \\ \vdots & \vdots & & \vdots \\ A_{q1} & A_{q2} & \cdots & A_{qq} \end{bmatrix},$$

where

$$A_{lm} = \Sigma_{j=1}^{N} (Z_{lj} - Z_{l.})(Z_{mj} - Z_{m.})^\dagger,$$

$NZ_{l.} = \Sigma_{j=1}^{N} Z_{lj}$, and Z_{ij} denotes jth independent observation on Z_i. The moments of λ_1, λ_2, λ_3, and λ_4 are also known to be given by

$$E\{\lambda_1^h\} = \left\{\prod_{j=1}^{s} \frac{\Gamma(n+h-j+1)}{\Gamma(n-j+1)}\right\} \left\{\prod_{i=1}^{q} \prod_{j=1}^{p_i} \frac{\Gamma(n-j+1)}{\Gamma(n+h-j+1)}\right\}, \tag{A7.5}$$

$$E\{\lambda_2^h\} = \frac{s^{hs}\,\Gamma(sn)}{\Gamma(sn+sh)} \prod_{j=1}^{s} \frac{\Gamma(n+h-j+1)}{\Gamma(n-j+1)}, \tag{A7.6}$$

$$E\{\lambda_3^h\} = \left(\frac{e}{n}\right)^{shn} \frac{|\Sigma_0|^{nh}}{|I + h\Sigma_0|^{n(1+h)}} \prod_{i=1}^{s} \frac{\Gamma(n+nh+1-i)}{\Gamma(n+1-i)}, \tag{A7.7}$$

$$E\{\lambda_4{}^h\} = \left(\frac{e}{n}\right)^{shn} \frac{1}{(1+h)^{sN(1+h)}} \prod_{i=1}^{s} \frac{\Gamma(N-i+Nh)}{\Gamma(N-i)}, \qquad \text{(A7.8)}$$

respectively.

Next, let us assume that $p_1 = \ldots = p_q = p$ and that $\Sigma_{ij} = 0$ $(i \neq j = 1, \ldots, q)$. Also, let H_5 denote the following hypothesis:

$$H_5 : \begin{cases} \Sigma_{11} & = \ldots = \Sigma_{q_1,q_1} \\ \Sigma_{q_1+1,q_1+1} & = \ldots = \Sigma_{q_2{}^*,q_2{}^*} \\ \vdots & \\ \Sigma_{q_{k-1}^*+1,q_{k-1}^*+1} & = \ldots = \Sigma_{q,q}, \end{cases}$$

where $q_0{}^* = 0, q_1{}^* = q_1, q_j{}^* = \Sigma_{i=1}^{j} q_i$ and $q_k{}^* = q$. We assume that N_i independent observations are available on $\boldsymbol{Z}_i$. In addition, let $N_i \boldsymbol{Z}_{i.} = \Sigma_{j=1}^{N_i} \boldsymbol{Z}_{ij}$ and

$$A_{ii} = \frac{1}{N_i} \sum_{j=1}^{N_i} (\boldsymbol{Z}_{ij} - \boldsymbol{Z}_{i.})(\boldsymbol{Z}_{ij} - \boldsymbol{Z}_{i.})^\dagger,$$

for $i = 1, \ldots, q$. The likelihood ratio test statistic λ_5 for H_5 and the moments of λ_5 are known to be as follows:

$$\lambda_5 = \frac{\Pi_{i=1}^{q} |A_{ii}/n_i|^{n_i}}{\Pi_{j=1}^{k} |\Sigma_{i=q_{j-1}^*+1}^{q_j{}^*} A_{ii}/n_i{}^*|^{n_j{}^*}}, \qquad \text{(A7.9)}$$

$$E\{\lambda_5{}^h\} = \left[\frac{\Pi_{i=1}^{k} n_i{}^{*phn_i{}^*}}{\Pi_{i=1}^{q} n_i^{phn_i}}\right] \prod_{i=1}^{p} \prod_{j=1}^{k} \left\{\left[\prod_{g=q_{j-1}^*+1}^{q_j{}^*} \frac{\Gamma(n_g + hn_g + 1 - i)}{\Gamma(n_g + 1 - i)}\right] \times \frac{\Gamma(n_j{}^* + 1 - i)}{\Gamma(n_j{}^* + hn_j{}^* + 1 - i)}\right\},$$

(A7.10)

where $n_i = N_i - 1$, and

$$n_j{}^* = \Sigma_{i=q_{j-1}^*+1}^{q_j{}^*} n_i.$$

Next, consider the hypothesis H_6,

$$H_6 : \begin{cases} \Sigma_{11} = \ldots = \Sigma_{qq} \\ \mu_1 = \ldots = \mu_q \\ \text{(under the assumption that } p_1 = \ldots = p_q = p \\ \text{ and } \Sigma_{ij} = 0 \text{ for } i \neq j). \end{cases}$$

The likelihood ratio test statistic λ_6 for H_6 and the moments of λ_6 are known to be given by

$$\lambda_6 = \frac{n_0^{pn_0} \Pi_{i=1}^q |A_{ii}|^{n_i}}{\Pi_{i=1}^q n_i^{pn_0} |\Sigma_{i=1}^q A_{ii} + \Sigma_{i=1}^q N_i(Z_{i.} - Z_{..})(Z_{i.}^\dagger - Z_{..}^\dagger)|^{n_0}},$$

and

$$E\left\{\lambda_6^h\right\} = \frac{n_0^{pn_0}}{\Pi_{i=1}^q n^{phn_i}} \prod_{i=1}^p \prod_{j=1}^q \left\{\frac{\Gamma(n_j + hn_j + 1 - i)}{\Gamma(n_j + 1 - i)}\right\} \frac{\Gamma(n_0 + q - i)}{\Gamma(n_0 + hn_0 + q - i)},$$

(using the same notation as in λ_5, and $n_0 = \Sigma_{i=1}^q n_i$).

The derivation of the likelihood ratio test statistics $\lambda_1, \lambda_2, \lambda_3, \lambda_4, \lambda_5$, and λ_6, and their moments, follow easily by following the same lines in the corresponding real cases.

Box (*130*) derived an asymptotic expression for the distribution function of a class of statistics W $(0 \leqslant W \leqslant 1)$ whose moments are of the form

$$E\{W^h\} = K\left(\prod_{j=1}^c y_j^{y_j}\right)\left(\prod_{k=1}^a x_k^{x_k}\right) \times \left\{\frac{\Pi_{k=1}^a \Gamma[x_k(1+h) + \xi_k]}{\Pi_{j=1}^c \Gamma[y_j(1+h) + \eta_j]}\right\},$$

$$h = 0, 1, \ldots \tag{A7.11}$$

where K is a normalizing constant such that $E\{W^0\} = 1$ and $\Sigma_{k=1}^a x_k = \Sigma_{j=1}^c y_j$.

Box also gave the first few terms only in the asymptotic expression. In several situations, the first few terms do not give the desired degree of accuracy. Using Box's method, Lee, Chang, and Krishnaiah (*131*) gave the terms up to the order of n^{-15}; these terms are linear combinations of the distribution functions of the central chi-square variates. The moments of λ_1, λ_2, λ_5, and λ_6 are of the form (A7.11). However, the asymptotic expression of Box is complicated if we have to take several terms in the series to get the desired degree of accuracy.

The distributions of certain powers of the statistics λ_1, λ_2, λ_3, λ_4, λ_5, and λ_6 are approximated in Krishnaiah, Lee, and Chang (*132*), Lee, Krishnaiah, and Chang (*133*), and Chang, Krishnaiah, and Lee (*134*), with Pearson's Type I distribution, by using the first four moments of these distributions. Using these approximations, they have also computed percentage points of the distributions of $\tilde{\lambda}_1$, $\tilde{\lambda}_2$, $\tilde{\lambda}_3$, $\tilde{\lambda}_4$, $\tilde{\lambda}_5$, and $\tilde{\lambda}_6$, where $\tilde{\lambda}_i = -2 \log \lambda_i$ for $i = 1, 2, \ldots, 6$. The accuracy of these approximations is sufficient for practical purposes. Nagarsenker and Das (*135*) have also computed the percentage points of the distribution of λ_2 for some values of the parameters by using a different method.

Khatri (*136*) derived the likelihood ratio test statistic for the reality of the covariance matrix of the complex multivariate normal population. One also can use other functions (like elementary symmetric funtions, ratios) of the roots of AA_1^{-1} for testing the hypothesis of the reality of the covariance matrix; here, A_1 denotes the real part of the sample *SP* matrix A. A certain power of the likelihood ratio statistic can be approximated with the Pearson's Type I distribution and the degree of accuracy of the approximation is sufficient for practical purposes.

In the next appendix, some applications of multivariate distributions will be given.

APPENDIX B
APPLICATIONS OF MULTIVARIATE DISTRIBUTIONS

B1. PRELIMINARIES

In this appendix, following Krishnaiah (*101*), we briefly discuss some applications of the complex multivariate distributions.

Let $X^t(t) = (X_1{}^t(t), \ldots, X_q{}^t(t))$ $(t = 1, \ldots, T)$ be a $1 \times u$ random vector that is distributed as a stationary Gaussian multivariate time series with zero mean vector and covariance matrix

$$R(\nu) = E\left\{ X(t)\, X^t(t+\nu) \right\}. \tag{B1.1}$$

Also, let the spectral density matrix

$$F(w) = \frac{1}{2\pi} \sum_{\nu=-\infty}^{\infty} e^{-i\nu w} R(\nu) \tag{B1.2}$$

be partitioned as

$$F(w) = \begin{bmatrix} F_{11}(w) & F_{12}(w) & \dots & F_{1q}(w) \\ F_{21}(w) & F_{22}(w) & \dots & F_{2q}(w) \\ \vdots & \vdots & & \vdots \\ F_{q1}(w) & F_{q2}(w) & \dots & F_{qq}(w) \end{bmatrix}, \tag{B1.3}$$

where $F_{jk}(w)$ is of order $p_j \times p_k$, and $X_j(t)$ is of order $p_j \times 1$. A well-known estimate [e.g., see Parzen (*137*), Brillinger (*95*)] of $F(w)$ is given by $\hat{F}(w) = (\hat{f}_{jk}(w))$, where

$$\hat{f}_{jk}(w) = \sum_{a=-m}^{m} w_a I_{jk}(w + (2\pi a/T)), \tag{B1.4}$$

$$I_{jk}(\lambda) = Z_j(\lambda) Z_k(\lambda), \tag{B1.5}$$

$$Z_j(\lambda) = \frac{1}{\sqrt{2\pi T}} \sum_{t=1}^{T} e^{-it\lambda} X_j(t). \tag{B1.6}$$

In the sequel, we assume that the weights w_a are equal to $1/(2m + 1)$.

It is known [see Goodman (*97*), Wahba (*138*), and Brillinger (*95*)] that $(2m + 1)\hat{F}(w)$ is approximately distributed as the central complex Wishart matrix with $(2m + 1)$ degrees of freedom.

B2. TESTING HYPOTHESES

We now discuss the problem of testing the hypothesis $H_1(w)$, where

$$H_1(w) : F_{jk}(w) = 0, \quad (j \neq k = 1, \dots, q). \tag{B2.1}$$

Let $s_i = \min(p_i, p_1 + \ldots + p_{i-1})$, and let $c_{i1} \leqslant \ldots \leqslant c_{is_i}$ be the eigenvalues of $\hat{\beta}_{ii}$, where

$$\hat{\beta}_{ii}(w) = \hat{F}_{ii}^{-1}(w)\,(\hat{F}_{i1}(w), \ldots, \hat{F}_{i,i-1}(w))$$

$$\begin{bmatrix} \hat{F}_{11}(w) & \ldots & \hat{F}_{1,i-1}(w) \\ \hat{F}_{21}(w) & \ldots & \hat{F}_{2,i-1}(w) \\ \vdots & & \vdots \\ \hat{F}_{i-1,1}(w) & \ldots & \hat{F}_{i-1,i-1}(w) \end{bmatrix}^{-1} \begin{bmatrix} \hat{F}_{1i}(w) \\ \\ \vdots \\ \hat{F}_{i-1,i}(w) \end{bmatrix}. \quad \text{(B2.2)}$$

The hypothesis $H_1(w)$ can be expressed as $H_1(w) = \cap_{j=2}^{q} H_{1j}(w)$, where $H_{1j}(w)$: $(F_{j1}(w), \ldots, F_{j,j-1}(w)) = 0$. We can test the hypothesis $H_1(w)$ as follows using a conditional approach. We first test the hypothesis $H_{12}(w)$. If $H_{12}(w)$ is rejected, we conclude that $H_1(w)$ is rejected. If $H_{12}(w)$ is accepted, we test $H_{13}(w)$, given $H_{12}(w)$. If $H_{13}(w)$ is rejected, we conclude that $H_1(w)$ is rejected; otherwise, we test $H_{14}(w)$ given, $H_{13}(w)$. This procedure is continued until a decision is made about the acceptance or rejection of $H_1(w)$.

Now, let $T_i(c_{i1}, \ldots, c_{is_i})$ denote a suitable function of $c_{i1}, \ldots, c_{is_i}$. Then the hypothesis H_{1j} given $\cap_{k=1}^{j-1} H_{1k}$ is accepted or rejected according as

$$T(c_{j1}, \ldots, c_{js_j}) \lesseqgtr d_j, \quad \text{(B2.3)}$$

for $j = 2, 3, \ldots, q$, with the understanding that H_{11} is equivalent to H_{12}. In Eq. (B2.3), the constants d_j are chosen such that

$$P[T(c_{j1}, \ldots, c_{js_j}) \leqslant d_j\,;\; j = 2, \ldots, q \mid H_1]$$

$$= \prod_{j=2}^{q} P[T(c_{j1}, \ldots, c_{js_j}) \leqslant d_j \mid H_1]$$

$$= (1 - \alpha). \quad \text{(B2.4)}$$

It is known that $(2m + 1)\,\hat{F}(w)$ is approximately distributed as the complex Wishart matrix with $(2m + 1)$ degrees of freedom, and $E(\hat{F}(w)) = (2m + 1)\,F(w)$.

Thus, when H_1 is true, the joint density of $c_{j1}, \ldots, c_{js_j}$ is approximately of the same form as Eq. (A3.12) after replacing $p, m,$ and q by s_j, $(2m + 1)$, and $\max(p_j, p_1 + \ldots + p_{j-1})$, respectively. When $T(c_{j1}, \ldots, c_{js_j}) = c_{js_j}$, the procedure discussed above is similar to the conditional approach used by Roy and Bargmann (*139*) for testing the multiple independence of several sets of variables, when their joint distribution is real multivariate normal. The test statistics $T(c_{j1}, \ldots, c_{js_j})$ can be also chosen to be equal to $\Pi_{i=1}^{s_j} (1 - c_{ji})$, c_{js_j}/c_{j1}, $c_{js_j}/\Sigma_{i=1}^{s_j} c_{ji}$, elementary symmetric functions of the roots, or some other suitable functions. When $q = 2$, analogous test statistics were used in the literature by various authors for testing the independence of two sets of variables when their joint distribution is real multivariate normal.

For example, test statistics analogous to $\Pi_{i=1}^{s_2} (1 - c_{2i})$ and $\Sigma_{i=1}^{s_2} c_{2i}$ were used by Wilks (*140*) and Bartlett (*141*), respectively, whereas test statistics analogous to c_{2s_2}/c_{21} and $c_{2s_2}/\Sigma_{i=1}^{s_2} c_{2i}$ were considered by Krishnaiah and Waikar (*142, 143*). In Eq. (B.4), one can of course use different types of statistics $T_j(c_{j1}, \ldots, c_{js_j})$ to test H_{1j}'s instead of using the same type of statistic $T(c_{j1}, \ldots, c_{js_j})$ at each stage of conditioning.

We now consider the problem of testing the hypothesis $H_2(w)$: $F_{11}(w) = \ldots = F_{qq}(w)$, when $R_{ij}(v) = R_{12}(v)$ $(i \neq j = 1, \ldots, q)$, and where $R_{ij}(v) = E[X_i(t)\,X_j^t(t + v)]$. Let

$$Y_1(t) = \{X_1(t) + \ldots + X_q(t)\}/q, \tag{B2.5}$$

and $Y_i(t) = X_i(t) - Y_1(t)$ for $i = 2, \ldots, q$. Then, $Y^t(t) = (Y_1^{\,t}(t), \ldots, Y_q^{\,t}(t))$ is a Gaussian stationary multiple time series with covariance matrix $R^*(v)$. The problem of testing the hypothesis $H_2(w)$ is equivalent to testing the hypothesis of $F_{12}^*(w) = 0$, where

$$F^*(w) = \begin{bmatrix} F_{11}^*(w) & F_{12}^*(w) \\ F_{21}^*(w) & F_{22}^*(w) \end{bmatrix} \tag{B2.6}$$

is the spectral density matrix of the time series $\{Y(t)\}$, and $F^*_{11}(w)$ is the spectral density matrix of the time series $\{Y_1(t)\}$. The hypothesis that $F^*_{12}(w) = 0$ can be tested by using the method described before. The method described above for testing $H_2(w)$ is analogous to the method used by Krishnaiah (*144*) for testing the equality of the diagonal blocks of the covariance matrix of the multivariate normal population when the off-diagonal blocks are equal.

Next, consider the problem of testing the hypothesis $H_3(w)$: $F(w) = F_0(w)$, where the matrix $F_0(w)$ is completely known. Let $c_p(w) \geqslant \ldots \geqslant c_1(w)$ be the latent roots of $\hat{F}(w)\, F_0^{-1}(w)$, and let $T(c_1(w), \ldots, c_p(w))$ be a suitable function of these roots. Also, let $\lambda_p(w) \geqslant \ldots \geqslant \lambda_1(w)$ be the latent roots of $\hat{F}(w)\, F_0^{-1}(w)$. Then the hypothesis $H_3(w)$, when tested against $T(\lambda_1(w), \ldots, \lambda_p(w)) > T(1, \ldots, 1)$, is accepted or rejected accordingly as

$$T(c_1(w), \ldots, c_p(w)) \lesseqgtr d_{3\alpha}\,, \tag{B2.7}$$

where

$$P[T(c_1(w), \ldots, c_p(w)) \leqslant d_{3\alpha} \mid H_3(w)] = (1 - \alpha). \tag{B2.8}$$

Likewise, one can apply a test procedure against two sided alternatives. When $H_3(w)$ is true, $\hat{F}(w)\, F_0^{-1}(w)$ is approximately distributed as the central complex Wishart matrix, with $(2m + - 1)$ degrees of freedom. Hence, the distributions of some of the statistics $T(c_1(w), \ldots, c_p(w))$ can be evaluated by using the results discussed in Appendix A.

Some possible choices of $T(c_1(w), \ldots, c_p(w))$, for instance, are $c_p(w)$, $\Sigma_{j=1}^p c_j(w)$, $c_p(w) - c_1(w)$, $\max_i(c_{i+1} - c_i)$, $c_p - \Sigma_{i=1}^p c_i/p$, or a statistic analogous to λ_3 of Section A7.

Another procedure for testing $H_3(w)$ against the alternative that $F(w) \neq F_0(w)$, is to accept $H_3(w)$ if

$$d_{3\alpha} \leqslant c_1(w) \leqslant c_p(w) \leqslant d_{4\alpha}\,, \tag{B2.9}$$

and to reject it otherwise, where

$$P[d_{3\alpha} \leqslant c_1(w) \leqslant c_p(w) \leqslant d_{4\alpha} \mid H_3(w)] = (1 - \alpha). \tag{B2.10}$$

The hypothesis $H_4(w) : F(w) = \sigma^2(w) I_p$ can be tested by using various ratios of the roots of $\hat{F}(w)$. For a review of these methods, the reader is referred to Krishnaiah and Schuurmann (*123*). Of course, one can test $H_4(w)$ also by using

$$\frac{s^s \, |\hat{F}(w) F_0^{-1}(w)|}{\{\mathrm{Tr}\, \hat{F}(w) F_0^{-1}(w)^{\,s}\}} \tag{B2.11}$$

as a test statistic. The distribution problems associated with this statistic were discussed in Section A7.

Now, let $\{X_1(t)\}, \ldots, \{X_q(t)\}$ be q independently distributed stationary, Gaussian p-variate time series, with spectral density matrices $F_1(w), \ldots, F_q(w)$. Also, let the record of the ith time series be T_i. In addition, let the sample estimate $\hat{F}_i(w)$ of $F_i(w)$ be defined in the same way as Eqs. (B1-4)-(B1.6) by taking the averages of $(2m_i + 1)$ periodograms, and let $c_{ijp}(w) \geq \ldots \geq c_{ij1}(w)$ be the roots of $\hat{F}_i(w) \hat{F}_j^{-1}(w)$, whereas $\lambda_{ijp} \geq \ldots \geq \lambda_{ij1}$ are the roots of $F_i(w) F_j^{-1}(w)$.

Also, let $\psi(\lambda_{ij1}, \ldots, \lambda_{ijp})$ be a suitable function of $\lambda_{ij1}, \ldots, \lambda_{ijp}$ with $\psi(1, \ldots, 1) = d$. We now discuss procedures for testing the hypothesis $H_5(w) : F_1(w) = \ldots = F_q(w)$. The hypothesis $H_5(w)$, when tested against

$$\cup_{i=1}^{q-1} \, [\psi(\lambda_{i,i+1,1}(w), \ldots, \lambda_{i,i+1,p}(w)) \geq d] \, , \tag{B2.12}$$

is accepted if

$$\psi(c_{i,i+1,1}(w), \ldots, c_{i,i+1,s}(w)) \leq c_\alpha \, , \tag{B2.13}$$

for $i = 1, \ldots, q - 1$, and rejected otherwise. Here c_α is chosen so as to satisfy

$$P[\psi(c_{i,i+1,1}(w), \ldots, c_{i,i+1,s}(w)) \leq c_\alpha \, ; \, i = 1, \ldots, q - 1 \, | H_5(w)]$$

$$= (1 - \alpha). \tag{B2.14}$$

When $H_5(w)$ is true, $\hat{F}_i(w) \hat{F}_j^{-1}(w)$ is distributed as the central complex multivariate F matrix, with $E(\hat{F}_i) = E(\hat{F}_j(w))$. Thus, one can use Bonferroni's inequality to compute bounds on the values of α in Eq. (B2.14).

Similarly, one can propose procedures against the alternatives

$$\cup_{i=1}^{q-1} [\psi(\lambda_{i,i+1,1}(w), \ldots, \lambda_{i,i+1,p}(w)) \leqslant d] \tag{B2.15}$$

and

$$\cup_{i=1}^{q-1} [\psi(\lambda_{i,i+1,1}(w), \ldots, \lambda_{i,i+1,p}(w)) \neq d] . \tag{B2.16}$$

We also can propose procedures for testing $H_5(w)$ against

$$\cup_{i=1}^{q-1} [\psi(\lambda_{iq1}, \ldots, \lambda_{iqp}) \neq d] , \tag{B2.17}$$

(or one-sided alternatives), by using $\psi(c_{iq1}, \ldots, c_{iqp})$ $(i = 1, \ldots, q - 1)$ as test statistics. If one tests $H_5(w)$ against

$$\cup_{i<j}^{q} [\psi(\lambda_{ij1}, \ldots, \lambda_{ijp}) \neq d] \tag{B2.18}$$

(or one-sided alternatives), one then uses $\psi(c_{ij1}, \ldots, c_{ijp})$ $(i < j = 1, \ldots, q)$ as test statistics.

The hypothesis $H_5(w)$ can be tested against

$$\cup_{i=1}^{q-1} [F_i(w) \neq F_{i+1}(w)] , \tag{B2.19}$$

$$\cup_{i=1}^{q-1} [F_i(w) \neq F_q(w)] , \tag{B2.20}$$

and

$$\cup_{i<j}^{q} [F_i(w) \neq F_j(w)] \tag{B2.21}$$

by using procedures analogous to those considered by Krishnaiah (*145*) and Krishnaiah and Pathak (*146*) for testing the equality of the covariance matrices of real multivariate normal populations. Procedures analogous to those considered for testing $H_5(w)$ can also be used for testing the hypothesis $F(w_1) = \ldots = F(w_k)$ against different alternatives, when $w_1, \ldots, w_k$ are widely separated, since, in this case, $\hat{F}(w_1), \ldots, \hat{F}(w_k)$ are distributed independently.

Next, consider the problem of testing the hypothesis

$$H_6(w) : F(w) = G_1(w) \otimes \Omega_1(w) + \ldots + G_k(w) \otimes \Omega_k(w), \tag{B2.22}$$

where $G_1(w), \ldots, G_k(w)$ are known matrices, $\Omega_1(w), \ldots, \Omega_k(w)$ are unknown matrices, and $\otimes$ denotes the Kronecker product. The hypothesis $H_6(w)$ can then be tested by using procedures analogous to those given by Krishnaiah and Lee (*147*) for testing the linear structures of the covariance matrices of real multivariate normal populations.

B3. DISCRIMINATION BETWEEN MULTIVARIATE NORMAL POPULATION

It is worthwhile mentioning that the techniques employed above in the area of the inference on multiple time series, are also useful in the area of the inference on the spectral density matrices of multivariate point processes.

We now discuss some procedures for testing the hypothesis on the adequacy of a given number of discriminators, to discriminate between complex multivariate normal populations [Krishnaiah (*101*)].

Let the rows of

$$Z_i = (z_{iju}) : m_i \times p(i = 1, \ldots, k) \tag{B3.1}$$

be distributed independently as complex multivariate normal, with mean vector μ_i^t and covariance matrix Σ. Also, we assume that $Z_1, \ldots, Z_k$ are distributed independently. The between group sums of squares and cross-products (*SP*) matrix, and the within group *SP* matrix, are respectively given by $S_1 = (s_{1uv})$, and $S_2 = (s_{2uv})$, where

$$s_{1uv} = \sum_{i=1}^{k} m_i(z_{i\cdot u} - z_{\cdot\cdot u})(\bar{z}_{i\cdot v} - \bar{z}_{\cdot\cdot v}), \tag{B3.2}$$

$$s_{2uv} = \sum_{i=1}^{k} \sum_{j=1}^{m_i} (z_{iju} - z_{i\cdot u})(\bar{z}_{ijv} - \bar{z}_{i\cdot v}), \tag{B3.3}$$

$$m_i z_{i\cdot u} = \Sigma_{j=1}^{m_i} z_{iju} \,, \tag{B3.4}$$

$$mz_{\cdot\cdot u} = \Sigma_{i=1}^{k} \Sigma_{j=1}^{m_i} z_{iju} \,, \tag{B3.5}$$

and

$$m = \Sigma_{i=1}^{k} m_i \,. \tag{B3.6}$$

We know that S_1 is distributed as the noncentral complex Wishart matrix, with $k - 1$ d.f., and

$$E[S_1/(k-1)] = \Sigma + [1/(k-1)] \, \nu\nu^t \,, \tag{B3.7}$$

where $\nu = (\mu_1 - \mu_{\cdot}, \ldots, \mu_k - \mu_{\cdot})$ and $k\mu = \Sigma_{j=1}^{k} \mu_j$. Also, S_2 is distributed as the central complex Wishart matrix, with $m - k$ degrees of freedom.

In the real case, likelihood ratio test for reducing the dimensionality, was discussed in the literature [see Rao (*148*)]. In the complex case, the analogous test was discussed by Young (*149*). Alternative procedures for the reduction of dimensionality are discussed below.

B4. ALTERNATIVE PROCEDURES

Let $l_p \geqslant \ldots \geqslant l_1$ denote the eigenvalues of $S_1 S_2^{-1}$, and let $\lambda_p \geqslant \ldots \geqslant \lambda_1$ be the eigenvalues of $\Omega = [1/(k-1)] \, \nu\nu^t \, \Sigma^{-1}$. Let $H_i : \lambda_i = 0$, and $A_i : \lambda_i > 0$. Then the nested hypotheses $H_1, \ldots, H_p$ can be tested simultaneously as follows.

We accept or reject H_i against A_i according to

$$l_i \lesseqgtr c_\alpha \,, \tag{B4.1}$$

where

$$P[l_p \leqslant c_\alpha \mid H_p] = (1 - \alpha) \,. \tag{B4.2}$$

The evaluation of the distribution of l_p in the central and noncentral cases was discussed in Appendix A.

Sometimes, the experimenter knows in advance that $\lambda_j > 0$ for $j = i + 1, \ldots, p$. Then, he has to test $H_1, \ldots, H_i$ only. In this case, the critical value c_α is chosen so as to satisfy

$$P[l_i \leqslant c_\alpha \mid H_i] = (1 - \alpha). \tag{B4.3}$$

But the distribution of l_i, under H_i, involves $\lambda_{i+1}, \ldots, \lambda_p$ as nuisance parameters. Thus, it is of interest to obtain bounds, free from nuisance parameters, on the probability integral in Eq. (B4.3).

Next, let $H_{ij} : \lambda_i = \lambda_j$, $A_{ij} : \lambda_i > \lambda_j$, and $f_{ij} = l_i/l_j$ for $i > j$. Then, the hypotheses H_{ij} $(i > j)$ can be tested simultaneously against A_{ij} as follows.

We accept or reject H_{ij} according to

$$f_{ij} \lesseqgtr d_\alpha , \tag{B4.4}$$

where

$$P[f_{p1} \leqslant d_\alpha \mid H_{p1}] = (1 - \alpha). \tag{B4.5}$$

The evaluation of the distribution of f_{p1} was discussed in Appendix A. When H_{p1} is true, the probability integral in Eq. (B4.5) involves λ_1 as a nuisance parameter. Thus, it is of interest to obtain a bound, free from nuisance parameters, on the probability integral in Eq. (B4.5). When $p > k - 1$, then $\lambda_1 = \ldots = \lambda_{p-k+1} = 0$, and so the probability integral in Eq. (B4.5) does not involve nuisance parameters. Also, in some practical situations, λ_1 is not significantly different from zero, and so it may be replaced by it.

The procedures discussed above are proposed in the same spirit as the procedures discussed by Krishnaiah and Waikar (*142, 143*) for the analogous real cases.

Next, consider the problem of testing the hypotheses $H_{t0}, \ldots, H_{q0}$ $(t \leqslant q \leqslant p)$, simultaneously against the alternatives $A_{t0}, \ldots, A_{q0}$, where $H_{j0} : \lambda_j \leqslant c \, \Sigma_{i=1}^{p} \lambda_i$ and $A_{j0} : \lambda_j > c \, \Sigma_{i=1}^{p} \lambda_i$ (c is a known constant). In this case, one accepts or rejects H_{j0} $(j = t, \ldots, q)$ according to

$$l_j/c \sum_{j=1}^{p} l_j \lesseqgtr d_\alpha , \tag{B4.6}$$

where

$$P[l_q/c \sum_{j=1}^{p} l_j \leqslant d_\alpha \mid H_{q0}] = (1 - \alpha). \tag{B4.7}$$

Similarly, we can test the hypothesis that

$$\sum_{i=t}^{q} \lambda_i < c \sum_{i=1}^{p} \lambda_i \tag{B4.8}$$

by using

$$\sum_{i=t}^{q} l_i/c \sum_{i=1}^{p} l_i \tag{B4.9}$$

as a test statistic.

The probability integrals, in both of the above cases, are not only complicated, but would involve nuisance parameters. Hence, it is of interest to obtain bounds, free from nuisance parameters, for these probability integrals. Similar tests can be used for drawing the inference on the eigenvalues of the spectral density matrix.

When Σ is known, one can use the above procedure after replacing $S_2/(m - k)$ by Σ. Procedures similar to those discussed above, for drawing inference on the eigenvalues of Ω, may be used for drawing inference on the canonical correlations, when the two sets of variables are jointly distributed as a complex multivariate normal.

In the methods discussed above for drawing inference on the spectral density matrices of the multiple time series, one may, of course, use alternative estimates $\hat{\hat{F}}(w)$ instead of using $\hat{F}(w)$ for estimating $F(w)$. When these estimates $\hat{\hat{F}}(w)$ are approximately distributed as complex Wishart matrices, then the distributions discussed in Appendices A and B are useful for computing the critical values.

Next, let us consider a matrix Σ^*, and let S^* be a suitable estimate of Σ^*. For example, in the area of principal component analysis, we may treat Σ^* and S^* as the population and sample covariance matrices, respectively. Similarly, in the area of canonical correlation analysis, we may treat Σ^* and S^* as the population canonical correlation matrix and sample canonical correlation matrix, respectively. Now, let $\theta_p \geqslant \ldots \geqslant \theta_1$ be the eigenvalues of S^*, whereas $\lambda_p \geqslant \ldots \geqslant \lambda_1$ denote the eigenvalues of Σ^*.

In some situations, the experimenter knows in advance that the λ_i's differ from each other. In these situations, he may be interested in simultaneous testing of the hypotheses H_{ij} against $A_{ij}(i > j)$, where $H_{ij} : \lambda_i < d\lambda_j$ $(d > 1)$ and $A_{ij} : \lambda_i > d\lambda_j$. In this case, we accept or reject H_{ij} according to

$$l_i/dl_j \lesseqgtr c_\alpha, \tag{B4.10}$$

where

$$P[l_p/l_1 \leqslant dc_\alpha \mid \lambda_p < d\lambda_1] = (1 - \alpha). \tag{B4.11}$$

A bound on the critical value c_α may be obtained by constructing a bound, free from nuisance parameters, on the left-hand side of Eq. (B4.11), and equating it to $(1 - \alpha)$.

One may, similarly, be interested in testing the hypotheses H_{ij} : $\lambda_i - \lambda_j < d$, $(d > 0)$ against the alternatives $A_{ij} : \lambda_i - \lambda_j > d$. In this case, one accepts or rejects H_{ij} against A_{ij} according to

$$(l_i - l_j - d) \lesseqgtr c_\alpha, \tag{B4.12}$$

where

$$P[l_p - l_1 \leqslant d + c_\alpha \mid \lambda_p - \lambda_1 < d] = (1 - \alpha). \tag{B4.13}$$

Here also, one may attempt to get a bound on c_α by constructing a lower bound, free from nuisance parameters, on the left-hand side of Eq. (B4.13).

Finally, we give the definitions of some complex multivariate processes. Let $\{X_j(t)\}$ $(j = 1, \ldots, n)$ denote n independent and identically distributed stationary p-variate continuous parameter $(-\infty < t < \infty)$ complex Gaussian stochastic process. Also, let $S(t) = X(t)\,X^\dagger(t)$, where

$$X(t) = [X_1(t), \ldots, X_u(t)]. \tag{B4.14}$$

The Hermitian matrix valued stochastic process $S(t)$, $(-\infty < t < \infty)$, obtained by varying t, is known [see Goodmann and Dubman (*150*)] to be as complex Wishart process. Now, let

$$S_1(t) = X(t) A X^\dagger(t) \tag{B4.15}$$

and

$$S_2(t) = X(t) X^\dagger(t) B , \tag{B4.16}$$

where $A : n \times n$ and $B : p \times p$ are symmetric matrices. The processes $\{S_1(t)\}$ and $\{S_2(t)\}$, $-\infty < t < \infty$, obtained by varying the time t, are generalizations of complex Wishart processes.

In the next appendix the ergodicity problem in the statistical theory of energy levels, will be discussed.

APPENDIX C
RANDOM MATRICES: ERGODIC PROPERTIES

In Chapters 5 and 7 it was mentioned that for transitions between the states of a given system, which is defined by a specific Hamiltonian, one usually relies on the ergodic behavior in making a comparison between theoretical ensemble results and experimental spectral results.

In this appendix we will discuss, following Brody, Flores, French, Mello, Pandey,and Wong (*42*), the problem of ergodicity. In order to justify the use of ensemble averages for dealing with the properties of single systems it must be shown that, with respect to the quantities which are compared (such as strength distributions, fluctuation measures, etc.), the ensemble has an appropriate ergodic behavior. Only very limited attention has been paid to this important question so far.

However, due to a recent work by Pandey (*151*), and by Pandey and French (*152*), the essential results for the conventional ensembles became available. Pandey and French show that a strong "locally generated" ergodicity can be obtained, and that the fluctuation measures, in particular, are stationary. The corresponding spectra are then closely analogous to the output of an ergodic stationary random process. Some attention will also be paid to ergodicity in reactions.

In the next section a preliminary introduction is given. Ergodicity for the one-point measures is discussed in Section C2, and for the two-point and higher-order correlation functions in Section C3. In Section C4 ergodicity and stationarity of transition strengths are discussed, whereas Section C5 is devoted to some remarks on the ergodicity of the S-matrix ensemble.

C1. PRELIMINARIES

In the text it was pointed out that an ensemble should have certain ergodic properties. We now examine these properties, and find out how far they might be applied to the random-matrix ensembles. Ergodic problems in conventional statistical mechanics were extensively studied [see, for example, Khinchin (*153*); Münster (*154*)]. A clear exposition of the basic concepts can be found, for instance, in the text of Balescu (*155*), and in Lebowitz and Penrose (*156*). For ergodicity in random processes (which is also more relevant to our topic) see, for instance, the texts of Yaglom (*157*), and Cramer and Leadbetter (*158*).

To begin with, let us consider a quantity $f(E, \xi)$, defined for H^{ξ}, a member of the ensemble, and dependent on the details of (namely a functional of) the spectrum in the neighborhood of E, say within the *measuring* interval $(E \pm \frac{1}{2}\delta E)$. We explicitly use the label ξ to emphasize that the quantities are defined for a particular member of the ensemble, but we shall often drop it for brevity. Triangular brackets (used in the sequel) will denote spectral averaging. $D(E)$ will denote the ensemble-averaged local spacing at E, which (as shown in the next section) can be replaced by a spectral-averaged local spacing. A spectral average is implied in the definition of some measures, as in the spacing variances. This is not the case, for instance, with $\Delta_3(\bar{n})$ when $\bar{n}$ covers the total observed spectrum. On the other hand, for much smaller $\bar{n}$ an experimental evaluation of the spectral average, given by Eq. (C1.1), may be carried out. This freedom, in fact, can be used to define sharper measures, as we shall see in Section C3 in the sequel. For instance, with $\Delta_3(\bar{n})$ we have $\delta E = \bar{n}D(E)$. Then f has a spectral average, over an *averaging* interval ΔE at E, given by [Brody *et al.* (*42*)] :

$$< f(E, \xi) >_p = \frac{1}{\Delta E} \int_{E-\frac{1}{2}\Delta E}^{E+\frac{1}{2}\Delta E} f(E', \xi)\, dE'$$

$$= \frac{1}{p} \int_{-p/2}^{p/2} f(E + rD, \xi)\, dr\,, \qquad \text{(C1.1)}$$

$$< f(E, \xi) >_p = \frac{1}{p} \sum_{E_i \epsilon \Delta} f(E_i, \xi). \qquad \text{(C1.2)}$$

In the above formulas the integral applies to a function defined for all the points on the energy axis, whereas the summation for one which exists only at the levels of E_i of the H^ξ matrix. The subscript p gives essentially the number of levels in the interval ΔE. (The precise definition will depend on the kind of averaging used, and the nature of the measure.)

In the last form of Eq. (C1.1) it was assumed, where necessary, that the spectrum has been unfolded. (Alternatively, one could, for local averaging, introduce the notion of a "tangent ensemble.") The function f also has an ensemble average given by

$$\bar{f}(E) = \int f(E, \xi)\, d\Phi\,(\xi)$$

$$= \int f(E, \xi)\, P(H^\xi) \prod_{i \leqslant j} dH_{ij}^\xi\,, \qquad \text{(C1.3)}$$

where $\Phi(\xi)$ is the distribution function for the ensemble. The corresponding density $P(H)$ for the Gaussian ensemble case was given in Chapter 9.

One could take the averaging interval ΔE to span the entire spectrum. In that case one would then have $p = d$. Ergodicity could then be defined as the equality of the two averages

$$< f(\xi) > = \overline{f}(E) \tag{C1.4}$$

in the $d \to \infty$ limit (which enters here just as does the infinite-time limit in ordinary statistical mechanics.)

Now Eq. (C1.4) is valid only if each of its two sides is equal to a constant, f_0, let us say, since they have different and independent arguments. Hence the formula separates into two conditions:

$$< f(\xi) > = f_0 = \overline{f}(E). \tag{C1.5}$$

The second of these formulas defines stationarity, whereas the first implies that the chosen H^ξ is a "characteristic" member of the ensemble. It will be noted that $f_0 = <\overline{f}>$, and that Eqs. (C1.4) and (C1.5) are valid almost always at best (since there will be exceptional members in the ensemble, and exceptional regions in the spectra).

However, averaging over the complete spectrum is not appropriate for our purposes, nor is it really necessary. Since only small segments of the spectrum are usually available experimentally, it follows that one must really study the results of *local*, rather than of *global*, spectral averaging. The definition of ergodicity, as stated by Eq. (C1.4), is then given in the limit $p \to \infty$, following $d \to \infty$ (keeping $p/d \to 0$). In practice, one would have $d \gg p \gg \delta E/D$. In most cases of interest (though not with proton resonances), the averaging interval ΔE will be small enough that, even without unfolding, the density can be taken as constant.

Accordingly, the main ergodicity requirement is the first equality of Eq. (C1.5), which can be restated as follows:

$$\mathrm{Var}_{(e)} < f(E, \xi) >_p \equiv \overline{< f(E, \xi) >_p^2} - \{\overline{< f(E, \xi) >_p}\}^2 \xrightarrow[p \to \infty]{} 0, \tag{C1.6}$$

where $\mathrm{Var}_{(e)}$ denotes the ensemble variance. [In the integral of Eq. (C1.1) it is to be understood that the measuring interval δE

moves continuously over the averaging interval. It is sometimes convenient to approximate the integral as a sum over a number, $\Delta E/\delta E$, of contiguous non-overlapping domains. The conditions given below for ergodic behavior apply equally well in both cases. The approximate procedure (which involves a new definition of spectral averaging) gives a variance (C1.6) larger than the integral method, but the difference is negligible when the number of intervals is large. The same holds for Eq. (C1.2).]

The ergodicity obtained via Eq. (C1.6) can be termed as "locally generated." (This only stresses the fact that the averaging interval can be a negligible fraction of the spectrum span.) On the other hand, the experimental data is available in the form of finite samples, and it becomes then imperative to know not only that the variance vanishes in the limit, but also its magnitude as a function of p. As for the second equality of Eq. (C1.5), *viz.*, stationarity over the averaging interval, it should be stressed that this is included in the definition of ergodicity merely to facilitate an easy interpretation of the physically-important quantity f_0 (to which the spectral average converges in the limit). In general, one may omit it, and then proceed with Eq. (C1.6) to verify ergodicity.

Except for "p-dependent" quantities (an example of which will be given at the end of the next section), we can assume that we may refer to as "local" stationarity. This is trivially achieved by local unfolding. Note that the local stationarity does not require or imply the global stationarity discussed in Section 5.3.

We now introduce the *autocovariance function* for f, defined by

$$S^f(E_1, E_2) = \overline{f(E_1)\, f(E_2)} - \bar{f}(E_1)\, \bar{f}(E_2). \tag{C1.7}$$

One can then take S^f to be locally stationary, and hence write it as a function of the relative coordinate r only. In the first case, Eq. (C1.1), we have $r = (E_2 - E_1)/D$, and in the second case, Eq. (C1.2), r is the number of levels in (E_1, E_2) with one of the ends included. $S^f(r)$ is an even function of r, and its absolute value is bounded by $S^f(0)$.

Accordingly, one has for Eq. (C1.1) and Eq. (C1.2), respectively, the following forms:

$$\mathrm{Var}_{(e)} < f(E) >_p = \frac{1}{p^2} \iint_{-p/2}^{p/2} S^f (E + r_1 D, E + r_2 D)\, dr_1\, dr_2$$

$$= \frac{2}{p^2} \int_0^p (p - r)\, S^f (r)\, dr, \qquad \text{(C1.8)}$$

$$\mathrm{Var}_{(e)} < f(E) >_p = \frac{1}{p^2} \sum_{E_i, E_j \in \Delta E} S^f (E_i, E_j)$$

$$= \frac{2}{p^2} \sum_{r=1}^{p} (p - r)\, S^f (r) + \frac{S^f (0)}{p}, \qquad \text{(C1.9)}$$

where we have integrated (or summed) over the center-of-mass variable. Since S^f determines the binary fluctuations of f (namely, fixes all its two-point measures), one sees that the fluctuations of f and its ergodic behavior are related to each other.

Equations (C1.8) and (C1.9) can now be applied either for a direct evaluation of the variance, or for inferring the asymptotic form of the variance, from that of S^f. Since S^f is already known for the one-point functions, we shall (in the next section) study their ergodicity by the first method. In other cases, the explicit evaluation of the variance could be a hard problem, and for these reasons one would rather use the second method.

A widely used result, giving a sufficient condition for the vanishing of the variance, is based on Slutsky's theorem [Slutsky (*159*); and Yaglom (*157*)]:

$$\frac{1}{p} \int_0^p S^f (r)\, dr \xrightarrow[p \to \infty]{} 0, \qquad \text{(C1.10)}$$

$$\frac{1}{p} \sum_{r=1}^{p-1} S^f (r) \xrightarrow[p \to \infty]{} 0, \qquad \text{(C1.11)}$$

An alternative version of the theorem (which is adequate for our purpose) is that the variance vanishes if $S^f (r)$ vanishes in the limit of large $|r|$, or is asymptotically a sinusoidal function of r. The theorem is valid for finite $S^f (0)$.

It turns out, in some cases, that $S^f(r)$ has delta-function singularities at $r = 0$. In these cases, we shall evaluate the integral over the singular part separately. One may also encounter functions which are anomalous in the sense that the variance given by Eq. (C1.6) vanishes for all p. This is actually an indication of an improper normalization of f. For example, if $f(E)$ has a proper ergodicity, namely one which is generated by taking p to be large enough, then $d^{-1} f(E)$ is such an anomalous function, and its ergodicity seems to be of no interest.

When the autocovariance function vanishes for all $r \neq 0$, as it does for transition strengths (which will be discussed in Section C4), the spectral average becomes a Gaussian random variable for large enough p. Likewise, one can show that the theorem of Diananda (*160*) leads to the same result when the autocovariance function vanishes for all $r > r_0$, where r_0 is a fixed number. One relies on a possible extension of this to argue that the kth-order spacing becomes Gaussian for large enough k.

The same argument then leads to the expectation that most of the ergodic measures should become Gaussian under spectral averaging for large enough p. Accordingly, for large but finite p, the variance will properly measure the statistical error. [When the distribution of $<f>_p$ is known, one can calculate confidence limits, like the probability that an observed value lies in the range $\bar{f} \pm \alpha(\mathrm{Var} <f>_p)^{1/2}$. It is in this sense that we refer to Var $<f>_p)^{1/2}$ as the statistical *error* of the estimation procedure. For nonzero $\bar{f}$ we shall also refer to the "*relative* error" $(\mathrm{Var} <f>_p)^{1/2}/\bar{f}$, whose square is the "figure of merit," as used by Dyson and Mehta (*54*).]

Unfortunately, the variance has not been calculated for most of the spectral-averaged two-point fluctuation measures. Therefore, the final comparison between theory and experiment is rather not conclusive.

C2. ONE-POINT MEASURES

The ergodic behavior of the level density was discussed by several authors. Grenander (*161*) proved that, for ensembles whose matrix elements have distributions which are symmetric about the zero, with identical variances and uniformly bounded moments of all finite order, the moments of the level density for almost all ensemble members go over, in the large-d limit, to those of the Wigner semi-circle distribution (see Chapter 4), in the sense that their variances about these values tend to zero. Grenander's results can then be extended to the other canonical ensembles [Brody *et al.* (*42*)].

Somewhat stronger results, but with less restrictions on the distributions of the matrix elements, were later on obtained by Arnold (*162*) and by Olson and Uppuluri (*163*). Using binary-association expansions, Mon and French (*164*) were then able to derive ergodicity also for the Gaussian orthogonal ensemble. No explicit spectral averaging appears in these calculations, though there is an implicit one, in the sense that results which are valid for moments of order much smaller than d are extended to all orders (thus generating a spectral smoothing which plays the same role).

We now return to the methods discussed in Section C1, and apply Eq. (C1.1) to the case where $f = \rho$. Since

$$d \int_{E-\frac{1}{2}\Delta E}^{E+\frac{1}{2}\Delta E} \rho(E')\, dE' \tag{C2.1}$$

counts the number of levels in ΔE, its variance, $\Sigma^2(p)$, can be formally calculated. One then obtains [Pandey (*151*)]:

$$\begin{aligned}\mathrm{Var}_{(e)} \langle \rho(E) \rangle_p &= \frac{1}{d^2(\Delta E)^2}\, \Sigma^2(p) \\ &= \left(\frac{\bar{\rho}(E)}{p}\right)^2 \Sigma^2(p).\end{aligned} \tag{C2.2}$$

This vanishes as $p^2 \ln p$ for the Gaussian ensembles, and as p^{-1} for the Poisson case [for which $\Sigma^2(p) = p$], yielding an ergodic behavior of the level density in all cases.

It will be noted that, because of the level repulsion (see Section 5.2), the convergence is faster for the Gaussian cases. In deriving ergodicity, we have used here the direct evaluation of the relevant variance instead of the general argument of the last section.

Ergodicity can also be shown for the number statistic, $n(E)$, namely the number of levels in an interval of fixed length $\bar{n}D$ at E.

It is then found that [Pandey (*151*)]:

$$\mathrm{Var}_{(e)} < n(E) >_p \; = \bar{n}^2 \left(\frac{\Sigma(p)}{p} \right)^2 , \tag{C2.3}$$

as long as the spectral averaging is done with $p/\bar{n}$ contiguous non-overlapping domains. It will be noted that the figure of merit for $< n(E) >$ is independent of $\bar{n}$, and therefore identical with that for $< \rho(E) >$ itself, as is seen from Eqs. (C2.2) and (C2.3), or could have been predicted from the fact that n is an additive statistic.

By the same token the figure of merit for the spectral-averaged kth-order spacing ($k < p$) is $\sigma^2(p - 1)/p^2$, which can be shown to differ only trivially from that for $< \rho >$ and $< n >$ for the Gaussian ensembles. [One has $(p + 1)$ levels in this case, and the spectral averaging is over $p/(k + 1)$ contiguous non-overlapping kth spacings.]

The application of the one-point measures lies in estimating the mean local spacing D. For example, D can be estimated in the obvious way, from the spectral-averaged spacings, with a fractional error $p^{-1}\sigma(p - 1)$. For $p = 100$, the fractional error is 10% for Poisson, but only 1.2%, 0.9% and 0.7% for the $\beta = 1, 2, 4$ ensembles (see Chapter 9). D can be estimated in other ways also; for example, as the spacing parameter for the "best" uniform spectrum.

One can then show that the estimator for D is given by the following formula:

$$A_{\min}(E; p) = \frac{p \sum_{i=1}^{p} iE_i - \sum_{i=1}^{p} i \sum_{i=1}^{p} E_i}{p \sum_{i=1}^{p} i^2 - \left(\sum_{i=1}^{p} i \right)^2}$$

$$= \frac{12}{p(p^2 - 1)} \sum_{i=1}^{p} \left(i - \frac{p+1}{2} \right) E_i , \tag{C2.4}$$

where the E_i are the energy levels (p of them) in ΔE. One can also show that $\bar{A}_{\min} = D$, and that

$$\mathrm{Var}_{(e)} A_{\min} = \frac{72\, D^2}{p^2 (p^2 - 1)^2} \sum_{i,j} \left(i - \frac{p+1}{2}\right)\left(\frac{p+1}{2} - j\right)$$

$$\sigma^2 (|i - j| - 1)$$

$$= \frac{12\, D^2}{p^2 (p^2 - 1)^2} \sum_{k=1}^{p-1} (-p^3 + 3p^2 k - 2k^3 + p - k)$$

$$\sigma^2 (k-1) \xrightarrow[\text{large } p]{} \frac{9\, D^2}{\beta \pi^2 p^2}, \qquad \text{(C2.5)}$$

where the last form is valid for the Gaussian ensembles.

The variance now being $\cong p^{-2}$ instead of $p^{-2} \ln p$, this estimator is asymptotically better than the simpler ones which were given before. Its variance, indeed, is quite close to the $(8D^2/\pi^2 p^2)$ of the optimal estimator for the Gaussian orthogonal ensemble given by Dyson and Mehta (*54*). For $p = 100$, however, the improvement over the simpler estimator is negligible, 1.0% instead of 1.2% for the Gaussian orthogonal ensemble.

Finally, it should be emphasized that not all the one-point measures are ergodic. For example, the levels in the fixed interval ΔE define the moments

$$M_\nu (E) = \frac{1}{p} \sum_{E_i \in \Delta E} E_i^\nu$$

$$= \frac{d}{p} \int_{E - \frac{1}{2} \Delta E}^{E + \frac{1}{2} \Delta E} E'^\nu \rho (E')\, dE'$$

$$= \frac{1}{p} \int_{-p/2}^{p/2} (E + rD)^\nu \frac{\rho (E + rD)}{\bar{\rho} (E + rD)}\, dr, \qquad \text{(C2.6)}$$

whose variance, for $\nu \geqslant 1$, diverge for asymptotic p. Thus, for $\nu = 1$, $\bar{M}_1 (E) = E$, and from the last formula we obtain:

$$\mathrm{Var}_{(e)}\, M_1(E) = \frac{1}{p^2} \iint_{-p/2}^{p/2} (E + r_1 D)(E + r_2 D)$$

$$[\delta(r_1 - r_2) - Y_2(r_1 - r_2)]\, dr_1\, dr_2$$

$$= \frac{2}{p^2} \int_0^p (p - r) \left(E^2 + \frac{p^2 D^2}{12} - \frac{prD^2}{6} - \frac{r^2 D^2}{6} \right)$$

$$[\delta(r) - Y_2(r)]\, dr$$

$$= D^2 \left[\frac{1}{p^2} \int_0^p r\, \Sigma^2(r)\, dr - \frac{\Sigma^2(p)}{4} \right]$$

$$+ E^2 \frac{\Sigma^2(p)}{p^2}\,. \qquad \text{(C2.7)}$$

Here, the second step follows by integrating over the center-of-mass variable, and the third step after two partial integrations. It will be noted that the first term in the last form of Eq. (C2.7) is of order $\ln p$ for the Gaussian ensembles, and is of order p for the Poisson case.

Similar results hold for the higher moments. These non-ergodic results are not surprising, however, since the levels remain highly correlated, even for large separations.

C3. HIGHER-ORDER FUNCTIONS

It is known that a k-point fluctuation measure can be expressed in terms of the k-point and lower-order correlation functions. As can be seen from Eq. (C1.7), the corresponding autocovariance function involves the $2k$-point function. The one-point measures could then be dealt with as in the preceding section, because the properties of the two-point function are well understood [Brody *et al.* (*42*)].

This is not the case, however, for the higher-order functions. Recently the ergodicity of a general k-point function was established

[Pandey (*151*)]. From this then follows the ergodicity of the fluctuation measures.

Accordingly, we define

$$T_k \equiv \left(\prod_{i=1}^{k} \bar{\rho}(x_i) \right)^{-1} \overline{S_k}$$

$$= \overline{\prod_{i=1}^{k} \{\rho(x_i)/\bar{\rho}(x_i)\}}, \tag{C3.1}$$

which are the k-point functions for unfolded spectra. Then T_k has delta-function singularities due to self-correlation of the levels, having a product of at most $(k - 1)$ delta functions in any term. T_1 is, of course, unity.

By ignoring all the self-correlation terms we get k-level correlation functions R_k, which are simply the joint-probability densities for k levels. (For a spectral realization of these functions see in the sequel.) Dyson's (*46*) k-level cluster functions Y_k are then obtained by subtracting out lower-order correlation effects from R_k. Formally one has

$$Y_k(x_1, x_2, \ldots, x_k) = \sum_G (-1)^{k-m} (m-1)! \prod_{j=1}^{m} R_{G_j}(x_t \text{ with } t \text{ in } G_j), \tag{C3.2}$$

where G denotes any division of the indices $(1, 2, \ldots, k)$ into unordered subsets $(G_1, G_2, \ldots, G_m)$. The inverse of Eq. (C3.2) is then given by

$$R_k(x_1, x_2, \ldots, x_k) = \sum_G (-1)^{k-m} \prod_{j=1}^{m} Y_{G_j}(x_t \text{ with } t \text{ in } G_j). \tag{C3.3}$$

Let us now take the k points (x_i) to be defined in a small segment of the spectrum, and write $x_i = x + r_i D(x)$, where x is the centroid, so that $\sum_{i=1}^{k} r_i = 0$. The functions Y_k are well defined and finite everywhere, being functions of the r_i and x. Moreover, the Y_k (and hence the T_k and the R_k) are globally stationary if they are independent of x, being a function then of the relative coordinates $r_{ij} \equiv (r_i - r_j)$ only.

The Poisson ensemble, which does not display any level repulsion (see Section 5.2), is by construction stationary for its cluster functions. In fact, $Y_k = 0$ for all $k > 1$. Dyson (*165*) has evaluated the Y_k for the circular ensembles and Mehta (*166*) for the Gaussian ones. In the large-d limit, the Gaussian ensembles give Y_k which are identical, at the center of the semicircle, with those for the circular ensembles.

It will be recalled that the circular ensembles are stationary by construction. Even so, they are inadequate for dealing with the physical problem of global stationarity (discussed in Section 5.3). This is so since the question here is whether or not such stationarity is obtained even when there is a large secular variation in the density.

It becomes, then, important to know whether or not the Gaussian ensembles give stationary results. (Ideally, one should make use of an embedded ensemble. However, since that is not tractable, one must continue to rely on the assumption that embedding is equivalent to a mapping which does not affect the fluctuation behavior.)

This question was discussed by Pandey (*151*) who has shown that, in the large-d limit, global stationarity is, in fact, obtained. From the stationarity of the Y_k follows that of all fluctuation measures [except for those, such as the moment variances given by Eq. (C2.7), in which a secular energy dependence is built into the definition].

Let us now come back to ergodicity. Consider the spectral average given by Eq. (C1.1), with $f = s_k$. (The dummy variable is the center-of-mass coordinate, x, in s_k.) For this we have the autocovariance function

$$S_k^{\rho} \equiv \overline{s_k(x; r_1, \ldots, r_k)\, s_k(x + rD; r_1, \ldots, r_k)} - \overline{s}_k(x; r_1, \ldots, r_k)\, \overline{s}_k(x + rD; r_1, \ldots, r_k), \qquad \text{(C3.4)}$$

in which the singularities (due to self-correlation of the levels) result from a product of at most $(2k - 1)$ delta functions.

However, as will be shown below by an example, the "observable" quantities are not s_k, but at least a $(k - 1)$-fold integral over s_k. This leaves us with only one delta-function involving r in S_k^ρ, which contributes a p^{-1} order term to the variance in Eq. (C1.8).

On the other hand, since the $(k > 1)$ cluster functions vanish whenever one of the relative coordinates r_{ij} is increased indefinitely, the continuous part of the S_k^ρ vanishes for all k in the limit of large r. In fact, one obtains [Pandey (*151*)]:

$$S_k^\rho(r) \xrightarrow[\text{large } r]{} 0(r^{-\alpha}), \tag{C3.5}$$

where $\alpha = 2$ for the orthogonal and unitary ensembles, and $\alpha = 1$ for the symplectic one. These considerations can be shown to yield the ergodic behavior of any k-point function.

We now consider the bounds for the errors involved in replacing one kind of average by another in the standard (ergodic) fluctuation measures. Suppose, then that for a quantity f our sample gives $\hat{p}$ contiguous non-overlapping measuring intervals [e.g., $\hat{p} = p/\bar{n}$ for $\Sigma^2(\bar{n})$]. Then one may assume that

$$\mathrm{Var}_{(e)} < f >_p \leqslant \frac{\mathrm{Var}_{(e)} f}{\hat{p}}, \tag{C3.6}$$

where the spectral averaging is done over the $\hat{p}$ non-overlapping intervals. [It is to be noted that the averaging over overlapping domains results in a smaller $\mathrm{Var}_{(e)} < f >_p$.] The upper bound follows by ignoring the correlations (assumed to be negative) between the values of f in distinct measuring intervals.

Since Eq. (C3.6) is an equality for the Poisson case, one can refer to the bound as the Poisson estimate. The inequality can be made plausible either by considering the asymptotic behavior of S_k^ρ, Eq. (C3.5), as done by Pandey (*151*), or by assuming that the spectral rigidity follows from negative correlations. Thus, for the number statistic the last form of Eq. (C2.3) is smaller than the Poisson estimate $\bar{n}\,\Sigma^2(\bar{n})/p$ by the factor $\bar{n}/p$ which, for practical purposes, is small.

The number variance Σ^2 is calculated from a given spectrum as $< (n - < n >)^2 >$. The Poisson estimate for the mean-square error in this case is, consequently, given by

$$(\bar{n}/p)\ \mathrm{Var}_{(e)}\ \left\{(n - \langle n \rangle)^2\right\},$$

which, with a Gaussian assumption for n, gives the figure of merit $(2\bar{n}/p)^{1/2}$. Similarly, for $\sigma^2(k)$ we find a Poisson figure of merit $[2(k+1)/p]^{1/2}$.

The ensemble variances for some measures were given by Dyson and Mehta. To take account of a further averaging over a set of measuring intervals, the same Poisson formula (C3.6) can be applied. For Δ_3, in particular, it is worth remarking that, whereas the standard experimental analysis makes use of the entire available spectrum as a measuring itnerval [thereby giving $\Delta_3(p)$], an alternative procedure would use a smaller measuring interval followed by an average over the intervals.

The Poisson estimate for $\langle \Delta_3(\bar{n}) \rangle$ then gives a figure of merit which can be minimized with respect to $\bar{n}$. For example, in the Gaussian ensemble by solving $\Delta_3(\bar{n}) = 2/\pi^2$, giving $\bar{n} \cong 8$ (independent of p). The general application of this sort of result would then be in seeking the best possible measures for comparing experiment with theory. Whether the Poisson estimation is sufficiently accurate, and whether the optimal Δ_3 is better than other measures, is unknown.

Since most of the standard fluctuation measures follow from the two-point function, one might also ask if the function itself can be estimated from the data and, if so, with what accuracy. In fact, this was done by Ideno and Ohkubo (*167*), and by Ideno (*168*), in order to search for "non-random distributions of neutron-resonance levels." They found stronger long-range correlations, resulting from periodic behavior in the spectrum. While the full implications of their results will not be drawn here, in the following we will nevertheless accept the essential parts of their procedure, and extend it to the general k-point case.

We start with a level at E_i and consider small intervals of length ϵD around $(k-1)$ points at distances $r_j D$, where $j = 1, 2, \ldots, k-1$, from E_i. The quantity of interest, say $\hat{R}_k(E_i; r_1, r_2, \ldots, r_{k-1})$, is the product of the numbers of levels in each interval. One calculates $\hat{R}_k$ for each level in the sample, and defines its spectral average by

$$\langle \hat{R}_k(E; r_1, r_2, \ldots, r_{k-1}) \rangle_p = \frac{1}{p} \sum_{E_i \text{ in } \Delta E} \hat{R}_k(E_i; r_1, r_2, \ldots, r_{k-1})$$

$$= \frac{d^{k-1}}{p} \sum_{E_i \text{ in } \Delta E} \prod_{j=1}^{k-1} \int_{(r_j - \epsilon/2)D}^{(r_j + \epsilon/2)D} \rho(E_i + y_j)\, dy_j$$

$$= \frac{1}{p \{\bar{\rho}(E)\}^k} \int_{-p/2}^{p/2} \rho(E + rD)\, dr \prod_{j=1}^{k-1} \int_{r_j - \epsilon/2}^{r_j + \epsilon/2} \rho(E + rD + r_j' D)\, dr_j', \qquad \text{(C3.7)}$$

where $|r_j| > \epsilon$ and $|r_j - r_i| > \epsilon$ for all i, j. Note that, because of trivial end effects, one cannot use the complete averaging interval [so that one should really write Eq. (C3.7) for $\hat{p} < p$ levels], nor can one permit any of the r_i to increase indefinitely.

We shall take ϵ to be small enough so that the last form in Eq. (C3.8) given below is a good approximation, but at the same time ϵ is large enough so that the figure of merit, resulting from Eq. (C3.9) given below, is within a reasonable limit. Typically we take $\epsilon = 0.1$. We then have

$$\overrightarrow{\langle \hat{R}_k(E; r_1, r_2, \ldots, r_{k-1}) \rangle_p} = \int_{r_1 - \epsilon/2}^{r_1 + \epsilon/2} \cdots \int_{r_{k-1} - \epsilon/2}^{r_{k-1} + \epsilon/2} R_k(E; r_1', r_2', \ldots, r_{k-1}')\, dr_1', \ldots, dr_{k-1}' \xrightarrow[\text{small } \epsilon]{} \epsilon^{k-1} R_k(E; r_1, r_2, \ldots, r_{k-1}), \qquad \text{(C3.8)}$$

so that $\epsilon^{-k+1} \langle \hat{R}_k \rangle$ gives a spectral realization of R_k.

It is easy to see that, for evaluating $\mathrm{Var}_{(e)} < \hat{R}_k >$ to the lowest order in ϵ, one must use the term with the maximum number of delta functions (k of them in this case, with the above restrictions on the r_i) in the autocovariance function of $\hat{R}_k$. One then finds

$$\mathrm{Var}_{(e)} < \hat{R}_k (E) >_p \;\rightarrow\; p^{-1}\, \epsilon^{k-1}\, R_k (E; r_1, r_2, \ldots, r_{k-1}), \tag{C3.9}$$

with the obvious ergodic property.

The relative error in the estimation procedure is then $(p\epsilon^{k-1} R_k)^{-1/2}$, which, for a given p, increases as k increases (because of its ϵ dependence). This makes plausible the proposition that a higher-than-two-point measure is not observable with any reasonable accuracy from the data.

It should be noted that the ergodic properties derived above for the three Gaussian ensembles are valid for energy levels having the same set of exact quantum numbers. Following Pandey (*151*) these ergodic properties can be extended to the mixed spectra, too. For a superposition of l independent ensembles, the cluster functions are then given by

$$Y_k (E; r_1, r_2, \ldots, r_k) = \sum_{i=1}^{l} (f_i)^k\, Y_{k,i} (E; f_i r_1, f_i r_2, \ldots, f_i r_k). \tag{C3.10}$$

Here the $Y_{k,i}$ are the functions for the ith-component ensemble, contributing on the average a fraction f_i of the energy levels to the mixed spectrum

$$\left(\sum_{i=1}^{l} f_i = 1 \right).$$

The nonsingular part of the autocovariance function, Eq. (C3.4), is a linear combination of the corresponding functions of the component ensembles. Accordingly, if all the component ensembles are ergodic for the k-point functions, the superposed ensemble is also ergodic.

It is clear from the above that the convergence (as $p \rightarrow \infty$) to ergodic results will, in general, be slower than the individual convergences. In fact, with $f_i \cong l^{-1}$, we have from Eq. (C3.10) for

$k > 1$ that $Y_k \to 0$ as $l \to \infty$, thus obtaining the Poisson ensemble as a limiting one. Equation (C3.6) can then be regarded as another demonstration of this rather well-known result.

C4. BEHAVIOR OF THE STRENGTH DISTRIBUTION

We now discuss the ergodic behavior of the strength distribution of the Gaussian ensemble. The Porter-Thomas distribution (see Chapter 7) in its finite (or asymptotic) form is an exact (or asymptotically exact) ensemble-averaged result. The question then arises as to whether or not the ensemble average equals (or approaches) the often measurable spectral average [Brody *et al.* (*42*)].

Ergodicity for eigenvector components was first considered by Brody and Mello (*169, 170*), who exploited a mathematical analogy between this problem and that of the ideal gas [whose ergodic properties, connected with fluctuations about the Maxwell-Boltzmann law, were already derived by Khinchin (*153*)]. We follow French and Pandey, however, in order to deal with the more general aspects of strength ergodicity.

Suppose then, for example, that the quantity of interest reduces to $x_{i\lambda}^2$, whose *ensemble* average is d^{-1}, where i denotes the eigenvector while λ denotes a basis vector in the statistical space, and both i and λ run from 1 to d. Since $\sum_i x_{i\lambda}^2 = 1$, we see that its *spectral* average is also d^{-1}.

Moreover, we know that in the Gaussian ensemble treatment of the (external) Porter-Thomas distribution, a change of basis reduces the strength to the $x_{i\lambda}^2$ form. Thus, for that function, we have a strict ergodicity, the two averages being exactly equal for every value of d. From the standpoint of the strength distributions, this merely tells us that the centroid can be calculated either as an ensemble or as spectral average.

This strict equality does not, of course, hold for higher moments. Consider, for instance, $x_{i\lambda}^4$. Its ensemble average is then $3/[d(d+2)]$. But if $\hat{\psi}_\lambda$ is an H eigenfunction (being identical with some Ψ_j), the spectral average is then d^{-1} [one state carrying unit "strength" and $(d-1)$ states carrying zero], which is the *largest* possible value of $x_{i\lambda}^4$. If, however, the $\hat{\psi}_\lambda$ in question were a uniform sum of all the H eigenvectors, we would find that $x_{i\lambda}^2$ would be d^{-1}

for every i, so that the spectral average of $x_{i\lambda}^4$ would be d^{-2}, which is its *smallest* possible value.

The ensemble average must, of course, lie between the two limits. Each of these states is quite "special." But the second, in which the strength is uniformly distributed, should be much less so than the first, in which it is completely concentrated. This follows from the fact that, for large d, the spectral value in the first case "misses" (larger than) the ensemble value, $3/d^2$, by a factor $d/3$, while in the second case it misses (smaller than) the ensemble value only by a factor of 3.

Let us now assume, as a characteristic case, which is adequate for discussing the Porter-Thomas distribution, that the quantity of interest is a function, say $\Omega(x_{i\lambda})$, of only a single component. We then do not restrict ourselves to averaging over the complete spectrum (d levels), but rather over a subset p of it only, as was done for the other quantities. (Alternatively, we can fix i and average over λ, and the results would be the same.)

Let us label the subset by $i = 1, 2, \ldots, p$, not necessarily implying any energy ordering by this labeling. The spectral average is now given by

$$< \Omega^{\xi} >_p \; = p^{-1} \sum_{i=1}^{p} \Omega^{\xi}(x_{i\lambda}). \qquad \text{(C4.1)}$$

The superscript ξ (which we usually omit), emphasizes that the value specifically depends on the H^{ξ} matrix. The question now is whether or not $< \Omega >_p$, for most members of the ensemble, approaches the ensemble-averaged value $\overline{\Omega}$.

Now the components are stationary, namely the ensemble distribution of $x_{i\lambda}$, and therefore $\overline{\Omega}(x_{i\lambda})$, do not depend on i (or λ). Hence $< \overline{\Omega} >_p = \overline{\Omega}$. The only condition that is left to be satisfied for ergodicity is the equivalent of Eq. (C1.6), namely,

$$\mathrm{Var}_{(e)} < \Omega >_p \; = p^{-1}\, \mathrm{Var}_{(e)}\, \Omega(x_{i\lambda}) + (1 - p^{-1})$$
$$\mathrm{Covar}_{(e)} \left\{ \Omega(x_{i\lambda}), \Omega(x_{j\lambda}) \right\} \xrightarrow[p \to \infty]{} 0. \qquad \text{(C4.2)}$$

Here Covar denotes the covariance of the two random variables in the argument. In deriving Eq. (C4.2), the first step follows from

Eq. (C1.9), and the fact that the variance is the same for all i, as is the covariance for all pairs $i \neq j$. Equation (C4.2) holds if and only if the covariance appearing in it vanishes. If, in addition, $\mathrm{Var}_{(e)}\, \Omega$ also vanishes, the (p-independent) ergodicity thus generated would be trivial (as was pointed out in Section C1).

Let us now restrict ourselves to the case where the function Ω is a polynomial. One can then show that

$$\mathrm{Var}_{(e)}\,(x_{i\lambda}^{\mu}) \cong d^{-\mu}\,, \tag{C4.3}$$

while $\mathrm{Covar}_{(e)}\left\{x_{i\lambda}^{\mu}, x_{j\lambda}^{\nu}\right\}$ vanishes identically unless μ, ν are both even, in which case it is of order $d^{-\frac{1}{2}(\mu+\nu)-1}$. To demonstrate non-trivial ergodicity, let us now renormalize our variables $x_{i\lambda}$ by defining the new variables

$$y_{i\lambda} = d^{\frac{1}{2}}\, x_{i\lambda}\,, \tag{C4.4}$$

and work with $\Omega(y_{i\lambda})$, where $\Omega(y)$ is independent of d.

In particular, one finds, with $\epsilon_{\mu} = 0, 1$, according to whether μ is odd or even, that

$$\begin{aligned} \mathrm{Covar}_{(e)}\left\{\langle y_{i\lambda}^{\mu}\rangle, \langle y_{i\lambda}^{\nu}\rangle\right\} = {} & \frac{\epsilon_{\mu+\nu}}{p}\, \{(\mu+\nu-1)!! \\ & - \epsilon_{\mu}\,\epsilon_{\nu}\,(\mu-1)!!\,(\nu-1)!!\} \\ & - \frac{\epsilon_{\mu}\,\epsilon_{\nu}}{2d}\, \mu\nu(\mu-1)!!\,(\nu-1)!! \\ & + O\left(\frac{1}{pd}, \frac{1}{d^2}\right), \end{aligned} \tag{C4.5}$$

which, term by term, corresponds to Eq. (C4.2). [Notice that for $\mu = \nu$ a slightly simpler form (asymptotically p^{-1}) emerges if one renormalizes each moment by its standard deviation.]

Since the second term on the right-hand side of Eq. (C4.5) vanishes as $d \to \infty$, it follows that

$$\mathrm{Var}_{(e)} \langle \Omega\,(y_{i\lambda})\rangle_p \cong p^{-1}\,, \tag{C4.6}$$

and hence one has an ergodic behavior. This behavior is valid also for polynomials in several components of the same or different vectors.

In Section C2 it was pointed out that ergodicity of the spectral moments leads to that of a spectrally-smoothed density function. The need for smoothing arises from the discrete nature of the Hamiltonian spectrum. One finds the same behavior with the $y_{i\lambda}$ distribution.

For a specified H one has for the spectral density function, with the renormalized amplitude $y_{i\lambda}$, the following formula:

$$g(y) = p^{-1} \sum_{i=1}^{p} \delta(y - y_{i\lambda}) . \tag{C4.7}$$

Since the ensemble average for $\delta(y - y_{i\lambda})$ is $\rho_G(y)$, namely a Gaussian density of zero centroid and unit variance, we also have $\overline{g}(y) = \rho_G(y)$. For the two-point function it follows then, from Eq. (C4.5), that

$$\text{Covar}_{(e)}\{g(y_1), g(y_2)\} = p^{-1} [\delta(y_1 - y_2)\, \rho_G(y_1) - \rho_G(y_1)\, \rho_G(y_2)] + O(d^{-1}). \tag{C4.8}$$

For integrated versions of $g(y)$, such as a histogram, Eq. (C4.8) then yields the expected p^{-1}-order variance. The same holds for strengths, namely the $y_{i\lambda}^2$ distribution.

Finally, it is worthwhile mentioning that in analyzing the data one often resorts to the maximum-likelihood procedure and, in order to test the Porter-Thomas distribution, estimates the number of degrees of freedom, ν_p, from a sample of size p. It is a property of such estimators as ν_p that, for large p, they become Gaussian, with a variance which vanishes in the limit. [For the general result see Kendall and Stuart (*171*) and, for the specific ν_p-result, see Porter and Thomas (*69*).] One could have, of course, predicted this from Eq. (C4.8), too.

C5. FINAL REMARKS

To conclude this appendix we briefly mention the ergodicity of the S-matrix ensembles. [For the S-matrix theory see Eden (*172*); Barut (*173*).]

The statistical theory of nuclear reactions sometimes makes use of ensembles of S-matrices also (see Section 4.5). The problem of ergodicity of the S-matrix ensembles was treated by Richert and Weidenmüller (*174*), who reduced the ergodicity of certain quantities to that of the widths and the energy levels of the ensemble of Hamiltonians under consideration. (The ergodicity of these quantities was taken for granted.)

The problem was also treated by French, Mello, and Pandey (*175*), and by Agassi, Weidenmüller, and Mantzouranis (*176*), who obtained the energy levels and widths from the eignevalues and eigenvectors of the matrices which belong to the Gaussian ensemble. (As was shown above, one has stationarity for this case.) We will not go through this topic, however, since it is beyond the scope of this book, and suggest that the reader will refer to the above mentioned references, to the early work of Coeseter (*41*), and to Brody *et al.* (*42*).

REFERENCES

1. L.I. Schiff, *Quantum Mechanics,* 2nd ed., McGraw-Hill, New York (1955).

2. L.S. Kisslinger and R.A. Sorensen, *Kgl. Danske Videnskab. Selskab. Mat-fys. Medd* 32, No. 9 (1960).

3. M. Baranger, Extension of the Shell Model for Heavy Spherical Nuclei, *Phys. Rev.* **120**, 957 (1960).

4. F.J. Dyson, Statistical Theory of the Energy Levels of Complex Systems. I, *J. Math. Phys.* 3, 140 (1962).

5. J.L. Rosen, J.S. Desjardins, J. Rainwater, and W.W. Havens, Jr., Slow Neutron Resonance Spectroscopy. I. U^{238}, *Phys. Rev.* 118, 687 (1960); Slow Neutron Resonance Spectroscopy. II. Ag, Au, Ta, *Phys. Rev.* **120**, 2214 (1960).

6. E.P. Wigner, Random Matrices in Physics, *SIAM Review* **9**, 1 (1967).

7. H. Goldstein, *Classical Mechanics,* Addison-Wesley, Cambridge, Massachusetts (1950).

8. B.O. Koopman, Hamiltonian Systems and Transformations in Hilbert Space, *Proc. Nat. Acad. Sci. U.S.A.* **17**, 315 (1931).

9. G.D. Birkhoff, Proof of a Recurrence Theorem for Strongly Transitive Systems, *Proc. Nat. Acad. Sci. U.S.A.* **17**, 650 (1931).

10. G.D. Birkhoff, Proof of the Ergodic Theorem, *Proc. Nat. Acad. Sci. U.S.A.* **17**, 656 (1931).

11. J. von Neumann, Proof of the Quasi-Ergodic Hypothesis, *Proc. Nat. Acad. Sci. U.S.A.* **18**, 70 (1932).

12. G.D. Birkhoff and B.O. Koopman, Recent Contributions to the Ergodic Theory, *Proc. Nat. Acad. Sci. U.S.A.* **18**, 279 (1932).

13. F. Ajzenberg and T. Lauritsen, Energy Levels of Light Nuclei, V, *Rev. Mod. Phys.* 27, 77 (1955).

14. E.P. Wigner, Statistical Properties of Real Symmetric Matrices with Many Dimensions, in *Can. Math. Congr. Proc.,* Univ. of Toronto Press, Toronto, Canada (1957), p. 174.

15. J.W. Mihelich, G. Scharff-Goldhaber, and M. McKeown, Decay Scheme of the 5.5-Hr Isomer of Hf^{180}, *Phys. Rev.* **94**, 794 (1954).

16. D.J. Hughes and J.A. Harvey, Neutron Cross Sections, Brookhaven National Laboratory Technical Report No. 325 (1955).

17. J.H. Christenson, J.W. Cronin, V.L. Fitch, and R. Turlay, Evidence for the 2π Decay of the $K_2{}^0$ Meson, *Phys. Rev. Letters* **13**, 138 (1964).

18. M. Carmeli, Statistical Theory of Energy Levels and Random Matrices in Physics, *J. Statist. Physics* **10**, 259 (1974).

19. C.E. Porter, Fluctuations of Quantal Spectra, in *Statistical Theories of Spectra: Fluctuations,* C.E. Porter, ed., Academic Press, New York (1965).

20. B. van der Pol, *Phil. Mag.* **26**(7), 921 (1938).

21. B. van der Pol and H. Bremmer, *Operational Calculus*, Cambridge University Press, London and New York (1959).

22. D.N. Lehmer, *List of Prime Numbers from 1 to 10,006,721*, Hafner, New York (1956).

23. L.D. Landau and E.M. Lifshitz, *Mechanics*, Pergamon Press, New York (1960).

24. E.P. Wigner, *Group Theory and its Applications to the Quantum Mechanics of Atomic Spectra*, Academic Press, New York (1959).

25. N. Rosenzweig, in *Statistical Physics,* K.W. Ford, ed., W.A. Benjamin, New York (1963).

26. A. Messiah, *Quantum Mechanics,* Vols. I and II, Wiley, New York (1961).

27. G.C. Wick, Invariance Principles of Nuclear Physics, *Ann. Rev. Nucl. Soc.* 8, 1 (1958).

28. H.A. Kramers, *Proc. Acad. Sci. Amsterdam* 33, 959 (1930).

29. M. Tinkham, *Group Theory and Quantum Mechanics,* McGraw-Hill, New York (1964).

30. K. Wilson, Proof of a Conjecture by Dyson, *J. Math. Phys.* 3, 1040 (1962).

31. L.K. Hua, *Am. J. Math.* 66, 470 (1944).

32. C. Chevalley, *Theory of Lie Groups,* Princeton University Press, Princeton, New Jersey (1946).

33. H. Weyl, *The Classical Groups,* Princeton University Press, Princeton, New Jersey (1946).

34. F.J. Dyson, The Threefold Way: Algebraic Structure of Symmetry Groups and Ensembles in Quantum Mechanics, *J. Math. Phys.* 3, 1199 (1962).

35. E.P. Wigner, On a Class of Analytic Functions from the Quantum Theory of Collisions, *Ann. Math.* 53, 36 (1951).

36. E.P. Wigner, Characteristic Vectors of Bordered Matrices with Infinite Dimensions, *Ann. Math.* 62, 548 (1955).

37. E.P. Wigner, Characteristic Vectors of Bordered Matrices with Infinite Dimensions II, *Ann. Math.* 65, 203 (1957).

38. E.P. Wigner, On the Distribution of the Roots of Certain Symmetric Matrices, *Ann. Math.* 67, 325 (1958).

39. C.E. Porter and N. Rosenzweig, Statistical Properties of Atomic and Nuclear Spectra, *Suomalaisen Tiedeakatemian Toimituksia* (*Ann. Acad. Sci. Fennicae*) AVI, No. 44 (1960); Repulsion of Energy Levels in Complex Atomic Spectra, *Phys. Rev.* 120, 1698 (1960).

40. B.V. Baronk, Accuracy of the Semicircle Approximation for the Density of Eigenvalues of Random Matrices, *J. Math. Phys.*, 5, 215 (1964).

41. F. Coeseter, The Symmetry of the *S* Matrix, *Phys. Rev.* 89, 619 (1953).

42. T.A. Brody, J. Flores, J.B. French, P.A. Mello, A. Pandey, and S.S.M. Wong, Random-Matrix Physics: Spectrum and Strength Fluctuations, *Revs. Mod. Phys.* 53, 385 (1981).

43. J.B. Grag, J. Rainwater, J.S. Petersen, and W.W. Havens, Jr., Neutron Resonance Spectroscopy, III. Th^{232} and U^{238}, *Phys. Rev.* 134, B985 (1964).

44. J. von Neumann and E.P. Wigner, *Phys. Z.* 30, 467 (1929).

45. L. Landau and Ya. Smorodinsky, *Lectures on the Theory of the Atomic Nucleus*, State Tech-Theoret. Lit. Press, Moscow (1955).

46. F.J. Dyson, Statistical Theory of the Energy Levels of Complex Systems, II, *J. Math. Phys.* 3, 157 (1962).

47. M. Gaudin, Sur la Loi Limite de l'Espacement des Valeurs Propes d'une Matrice Aleatoire, *Nucl. Phys.* 25, 447 (1961).

48. E.P. Wigner, *Conference on Neutron Physics by Time of Flight* (Gatlinburg, Tennessee, November 1956), Oak Ridge National Lab. Report ORNL-2309 (1957).

49. M.L. Mehta, On the Statistical Properties of the Level-spacings in Nuclear Spectra, *Nucl. Phys.* 18, 395 (1960).

50. M.L. Mehta and M. Gaudin, On the Density of Eigenvalues of a Random Matrix, *Nucl. Phys.* 18, 420 (1960).

51. F.J. Dyson, Statistical Theory of Energy Levels of Complex Systems, III, *J. Math. Phys.* 3, 166 (1962).

52. P.B. Kahn, Energy Level Spacing Distributions, *Nucl. Phys.* 41, 159 (1963).

53. C.E. Porter, Random Matrix Diagonalization – Some Numerical Computations, *J. Math. Phys.* 4, 1039 (1963).

54. F.J. Dyson and M.L. Mehta, Statistical Theory of the Energy Levels of Complex Systems, IV, *J. Math. Phys.* 4, 701 (1963).

55. J. Gunson, Proof of a Conjecture by Dyson in the Statistical Theory of Energy Levels, *J. Math. Phys.* 3, 752 (1962).

56. M.L. Mehta and F.J. Dyson, Statistical Theory of the Energy Levels of Complex Systems, V, *J. Math. Phys.* 4, 713 (1963).

57. C.E. Porter, Further Remarks on Energy Level Spacings, *Nucl. Phys.* 40, 167 (1963).

58. P.B. Kahn and C.E. Porter, Statistical Fluctuations of Energy Levels: The Unitary Ensemble, *Nucl. Phys.* 48, 385 (1963).

59. H.S. Leff, Systematic Characterization of *m*th-Order Energy Level Spacing Distributions, *J. Math. Phys.* 5, 756 (1964).

60. H.S. Leff, Class of Ensembles in the Statistical Theory of Energy Level Spectra, *J. Math. Phys.* 5, 763 (1964).

61. D. Fox and P.B. Kahn, Identity of the *n*th-Order Spacing Distributions for a Class of Hamiltonian, *Phys. Rev.* 134, B1151 (1964).

62. I. Gurevich and M.I. Pevsher, Repulsion of Nuclear Levels, *Nucl. Phys.* 2, 575 (1956).

63. A.M. Lane, in *Conference on Neutron Physics by Time of Flight* (Gatlinburg, Tennessee, November 1956), Oak Ridge National Lab. Report ORNL-2309 (1957).

64. R.E. Trees, Repulsion of Energy Levels in Complex Atomic Spectra, *Phys. Rev.* 123, 1293 (1961).

65. L. Dresner, Spacings of Nuclear Energy Levels, *Phys. Rev.* 113, 633 (1959).

66. F.J. Dyson, A Brownian Motion Model for the Eigenvalues of a Random Matrix, *J. Math. Phys.* 3, 1191 (1962).

67. L.D. Favro, P.B. Khan, and M.L. Mehta, Concerning Polynomials Encountered in the Study of the Distribution Function of

Spacings, Brookhaven National Lab. Report No. BNL-757 (T-280), September 1932.

68. J.M. Scott, *Phil. Mag.* **45**, 1322 (1952).

69. C.E. Porter and R.G. Thomas, Fluctuations of Nuclear Reaction Widths, *Phys. Rev.* **104**, 483 (1956).

70. J.A. Harvey, D.J. Hughes, R.S. Carter, and V.E. Pilcher, Spacings and Neutron Widths of Nuclear Energy Levels, *Phys. Rev.* **99**, 10 (1955).

71. D.J. Hughes and J.A. Harvey, Size Distribution of Neutron Widths, *Phys. Rev.* **99**, 1032 (1955).

72. C.E. Porter, Statistics of Atomic Radiative Transition Probabilities, *Phys. Letters* 2, 292 (1962).

73. N. Ullah and C.E. Porter, Expectation Value Fluctuations in the Unitary Ensemble, *Phys. Rev.* **132**, 948 (1963).

74. N. Ullah and C.E. Porter, Invariance Hypothesis and Hamiltonian Matrix Elements Correlations, *Phys. Letters* **6**, 301 (1963).

75. N. Rosenzweig, Anomalous Statistics of Partial Radiation Widths, *Phys. Letters* **6**, 123 (1963).

76. L.K. Hua, Harmonic Analysis of Functions of Several Complex Variables in the Classical Domains, in *Translations of Mathematical Monographs,* Vol. 6, Am. Math Soc., Providence, Rhode Island (1963); L.K. Hua, *American Mathematical Society Translations,* Vols. 32 and 33, Am. Math. Soc., Providence, Rhode Island (1963).

77. M. Carmeli and S. Malin, *Representations of the Rotation and Lorentz Groups,* Dekker, New York (1976).

78. J.B. Grag, *Statistical Properties of Nuclei*, Plenum Press, New York (1972).

79. E.P. Wigner, Introductory Talk, in Ref. 78, p. 7.

80. G. Joos, *Introduction to Theoretical Physics,* Hafner Publication Co., New York (1958).

81. J.B. French and S.S. Wong, Validity of Random Matrix Theories for Many-Particle Systems, *Phys. Letters* **33B**, 449 (1970).

82. O. Bohigas and J. Flores, Two-Body Random Hamiltonian and Level Density, *Phys. Letters* **34B**, 261 (1971).

83. J.B. French and F.S. Chang, Distribution Theory of Nuclear Level Densities and Related Quantities, Ref. 78, p. 405.

84. O. Bohigas and J. Flores, Some Properties of Level Spacing Distributions, Ref. 78, p. 195.

85. R. Balian, Random Matrices and Information Theory, *Nuovo Cimento* **B57**, 183 (1968).

86. S.N. Roy, *Some Aspects of Multivariate Analysis*, Wiley (1957).

87. A.T. James, Distributions of Matrix Variates and Latent Roots Derived from Normal Samples, *Ann. Math. Stat.* **35**. 475 (1964).

88. P.R. Krishnaiah and T.C. Chang, On the Exact Distributions of the Extreme Roots of the Wishart and Manova Matrices, *J. Multivariate Analysis* **1**, 108 (1971).

89. P.R. Krishnaiah and V.B. Waikar, Exact Distributions of Any Few Ordered Roots of a Class of Random Matrices, *J. Multivariate Analysis* **1**, 308 (1971).

90. P.R. Krishnaiah and A.K. Chattopadhyay, On Some Noncentral Multivariate Distributions, *South African Statist. J.* **9**, 37 (1975).

91. E.J. Hannan, *Multiple Time Series,* Wiley, New York (1970).

92. W.S. Liggett, Jr., Passive Sonar: Fitting Models to Multiple Time Series, in *Proceedings of the NATO Advanced Study Institute on Signal Processing,* Academic Press, New York (1972).

93. W.S. Liggett, Jr., Determination of Smoothing for Spectral Matrix Estimation, Technical Report, Raytheon Company, Portsmouth, Rhode Island (1973).

94. M.B. Priestley, T. Subba Rao, and H. Tong, Identification of the Structure of Multivariable Stochastic Systems, in *Multivariate Analysis, III,* P.R. Krishnaiah, Ed., Academic Press, New York (1973).

95. D.R. Brillinger, *Time Series: Data Analysis and Theory*, Holt, Rinehart and Winston, New York (1974).

96. R.A. Wooding, The Multivariate Distribution of Complex Normal Variables, *Biometrika* **43**, 212 (1956).

97. N.R. Goodman, Statistical Analysis Based on a Certain Multivariate Complex Gaussian Distribution (An Introduction), *Ann. Math. Statist.* **34**, 152 (1963).

98. A.T. James, Distributions of Matrix Variates and Latent Roots Derived from Normal Samples, *Ann. Math. Statist.* **35**, 475 (1964).

99. E.P. Wigner, Distribution Laws for the Roots of a Random Hermitian Matrix, in *Statistical Theories of Spectra: Fluctuations,* C.E. Porter, Ed., Academic Press, New York (1965).

100. C.G. Khatri, Classical Statistical Analysis Based on a Certain Multivariate Complex Gaussian Distribution, *Ann. Math. Statist.* **36**, 98 (1965).

101. P.R. Krishnaiah, Some Recent Developments on Complex Multivariate Distributions, *J. Multivariate Anal.* **6**, 1 (1976).

102. C. Andreief, Note Sur Une Relation Entre les Integrales Defines des Produits des Functions, *Mem. Soc. Sci. Bordeaux* 3(2), 1 (1883).

103. C.G. Khatri, On the Moments of Traces of Two Matrices in Three Situations for Complex Multivariate Normal Populations, *Sankhya Ser. A* 32, 65 (1970).

104. L.K. Hua, *Harmonic Analysis of Functions of Several Complex Variables in the Complex Domains* (in Russian), Moscow (1959).

105. A.T. James, Zonal Polynomials of the Real Positive Definite Symmetric Matrices, *Ann. Math.* 74, 456 (1961).

106. C.S. Herz, Bessel Functions of Matrix Argument, *Ann. Math.* 61, 474 (1955).

107. M.S. Srivastava, On the Complex Wishart Distribution, *Ann. Math. Statist.* 36, 313 (1965).

108. V.B. Waikar, T.C. Chang, and P.R. Krishnaiah, Exact Distributions of Few Arbitrary Roots of Some Complex Random Matrices, *Austral. J. Statist.* 14, 84 (1972).

109. D.G. Kabe, Complex Analogues of Some Classical Noncentral Multivariate Distributions, *Austrl. J. Statist.* 8, 99 (1966).

110. G.L. Turin, The Characteristic Function of Hermitian Quadratic Forms in Complex Normal Variables, *Biometrika* 43, 199 (1960).

111. C.G. Khatri, Distribution of the Largest or the Smallest Characteristic Root Under Null Hypothesis Concerning Complex Multivariate Normal Populations, *Ann. Math. Statist.* 35, 1807 (1964).

112. S. Al-Ani, On the Distribution of the *i*th Latent Root Under Null Hypotheses Concerning Complex Multivariate Normal

Populations, Mimeo Ser. No. 145, Dept. of Statistics, Purdue University; *Canad. Math. Bull.* **15**, 321 (1968).

113. C.G. Khatri, Noncentral Distributions of *i*th Largest Characteristic Roots of Three Matrices Concerning Complex Multivariate Normal Populations, *Ann. Inst. Statist. Math.* **21**, 23 (1969).

114. K.C.S. Pillai and D.L. Young, An Approximation to the Distribution of the Largest Root of a Complex Wishart Matrix, *Ann. Inst. Statist. Math.* 22, 89 (1970).

115. K.C.S. Pillai and G.M. Jouris, An Approximation to the Distribution of the Largest Root of a Matrix in the Complex Gaussian Case, *Ann. Inst. Statist. Math.* **24**, 61 (1972).

116. F.J. Schuurmann and V.B. Waikar, Upper Percentage Points of the Individual Roots of the Complex Wishart Matrix, *Sankhya Ser. B* (1974).

117. P.R. Krishnaiah and F.J. Schuurmann, On the Exact Distributions of the Individual Roots of the Complex Wishart and Multivariate Beta Matrices, ARL 74-0063, Aerospace Research Laboratories, Wright-Patterson AFB, Ohio (1974).

118. P.R. Krishnaiah and F.J. Schuurmann, On the Exact Joint Distributions of the Extreme Roots of the Complex Wishart and Multivariate Beta Matrices, ARL 74-0111, Aerospace Research Laboratories, Wright-Patterson AFB, Ohio (1974).

119. M.L. Mehta, *Random Matrices,* Academic Press, New York (1967).

120. K.C.S. Pillai and G.M. Jouris, Some Distribution Problems in the Multivariate Complex Gaussian Case, *Ann. Math. Statist.* **42**, 517 (1971).

121. P.R. Krishnaiah and F.J. Schuurmann, On the Exact Distribution of the Trace of a Complex Multivariate Beta Matrix, ARL 74-0107, Aerospace Research Laboratories, Wright-Patterson AFB, Ohio (1974).

122. P.R. Krishnaiah and F.J. Schuurmann, Approximations to the Traces of Complex Multivariate Beta and F Matrices, ARL 74-0123, Aerospace Research Laboratories, Wright-Patterson AFB, Ohio (1974).

123. P.R. Krishnaiah and F.J. Schuurmann, On the Evaluation of Some Distributions that Arise in Simultaneous Tests for the Equality of the Latent Roots of the Covariance Matrix, *J. Multivariate Anal.* **4**, 265 (1974).

124. P.R. Krishnaiah and F.J. Schuurmann, On the Distributions of the Ratios of the Extreme Roots of the Real and Complex Multivariate Beta Matrices, ARL 74-0122, Aerospace Research Laboratories, Wright-Patterson AFB, Ohio (1974).

125. N.R. Goodman, The Distribution of the Determinant of a Complex Wishart Distributed Matrix, *Ann. Math. Statist.* **34**, 178 (1963).

126. N. Giri, On the Complex Analogues of T^2 and R^2 Tests, *Ann. Math. Statist.* **36**, 664 (1965).

127. G. Wahba, Some Tests of Independence for Stationary Multivariate Time Series, *J. Roy. Statist. Soc. Ser. B* **33**, 153 (1971).

128. K.C.S. Pillai and B.N. Nagarsenker, On the Distribution of the Sphericity Test Criterion in Classical and Complex Normal Populations Having Unknown Covariance Matrices, *Ann. Math. Statist.* **42**, 764 (1971).

129. A.K. Gupta, Distribution of Wilk's Likelihood-Ratio Criterion in the Complex Case, *Ann. Inst. Statist. Math.* **23**, 77 (1971).

130. G.E.P. Box, A General Distribution Theory for a Class of Likelihood Criteria, *Biometrika* **36**, 317 (1949).

131. J.C. Lee, T.C. Chang, and P.R. Krishnaiah, Approximations to the Distributions of the Likelihood Ratio Statistics for Testing Certain Structure on the Covariance Matrices of Real Multi-

variate Normal Populations, ARL 75-0167, Aerospace Research Laboratories, Wright-Patterson AFB, Ohio (1975).

132. P.R. Krishnaiah, J.C. Lee, and T.C. Chang, Approximations to the Distributions of the Likelihood Ratio Statistics for Testing Certain Structures on the Covariance Matrices of Complex Multivariate Normal Populations, ARL 75-0169, Aerospace Research Laboratories, Wright-Patterson AFB, Ohio (1975).

133. J.C. Lee, P.R. Krishnaiah, and T.C. Chang, Approximations to the Distributions of the Determinants of Real and Complex Multivariate Beta Matrices, ARL 75-0168, Aerospace Research Laboratores, Wright-Patterson AFB, Ohio (1975).

134. T.C. Chang, P.R. Krishnaiah, and J.C. Lee, Approximations to the Distributions of the Likelihood Ratio Statistics for Testing the Hypotheses on Covariance Matrices and Mean Vectors Simultaneously, ARL 75-0176, Aerospace Research Laboratories, Wright-Patterson AFB, Ohio (1975).

135. B.N. Nagarsenker and M.M. Das, Exact Distribution of Sphericity Criterion in the Complex Case and Its Percentage Points, *Communications in Statistics* 4, 363 (1975).

136. C.G. Khatri, A Test for Reality of a Covariance Matrix in a Certain Complex Gaussian Distribution, *Ann. Math. Statist.* 36, 115 (1965).

137. E. Parzen, Multiple Time Series Modelling, in *Multivariate Analysis, II,* P.R. Krishnaiah, Ed., Academic Press, New York (1969).

138. G. Wahba, On the Distributions of Some Statistics Useful in the Analysis of Jointly Stationary Time Series, *Ann. Math. Statist.* 39, 1849 (1968).

139. S.N. Roy and R. Bargmann, Tests of Multiple Independence and the Associated Confidence Bounds. *Ann. Math. Statist.* 29, 491 (1958).

140. S.S. Wilks, On the Independence of K Sets of Normally Distributed Statistical Variables, *Econometrica* 3, 309 (1935).

141. M.S. Bartlett, A Note on Tests of Significance in Multivariate Analysis, *Proc. Camb. Phil. Soc.* **35**, 180 (1939).

142. P.R. Krishnaiah and V.B. Waikar, Simultaneous Tests for Equality of Latent Roots Against Certain Alternatives, I, *Ann. Inst. Statist. Math.* 23, 451 (1971).

143. P.R. Krishnaiah and V.B. Waikar, Simultaneous Tests for Equality of Latent Roots Against Certain Alternatives, II, *Ann. Inst. Statist. Math.* **24**, 81 (1972).

144. P.R. Krishnaiah, Tests for the Equality of the Covariance Matrices of Correlated Multivariate Normal Populations, in *A Survey of Statist. Design Linear Models*, J.N. Srivastava, Ed., North-Holland (1975).

145. P.R. Krishnaiah, Simultaneous Tests for the Equality of Covariance Matrices Against Certain Alternatives, *Ann. Math. Statist.* **39**, 1303 (1968).

146. P.R. Krishnaiah and P.K. Pathak, A Note on Confidence Bounds for Certain Ratios of Characteristic Roots of Covariance Matrices, *Austrl. J. Statist.* **10**, 116 (1968).

147. P.R. Krishnaiah and J.C. Lee, On Covariance Structures, *Sankhya*, to appear; also see ARL 75-0103 (1976).

148. C.R. Rao, *Linear Statistical Inference and Its Applications*, Wiley, New York (1964).

149. J.C. Young, Some Inference Problems Associated with the Complex Multivariate Normal Distribution, Technical Rept. No. 102, Dept. of Statistics, Southern Methodist University (1971).

150. N.R. Goodman and M.R. Dubman, Theory of Time-Varying Spectral Analysis and Complex Wishart Matrix Processes, in

Multivariate Analysis, II, P.R. Krishnaiah, Ed., Academic Press, New York (1969).

151. A. Pandey, Doctoral Dissertation, University of Rochester (1978); Statistical Properties of Many-Particle Spectra, III. Ergodic Behavior in Random-Matrix Ensembles, *Ann. Phys.* (N.Y.) **119**, 170 (1979).

152. A. Pandey and J.B. French, Binary Correlations in Random Matrix Spectra, *J. Phys.* **A12**, L83 (1979).

153. A.L. Khinchin, *Mathematical Foundations of Statistical Mechanics*, Dover, New York (1949).

154. A. Münster, Prinzipien der Statistischen Mechanik, in *Prinzipien der Thermodynamik und Statistik*, Handbuch der Physik, Ed. S. Flügge, Springer, Berlin (1959).

155. R. Balescu, *Equilibrium and Non-Equilibrium Statistical Mechanics,* Wiley-Interscience, N.Y. (1975).

156. J.L. Lebowitz and O. Penrose, Modern Ergodic Theories, *Phys. Today* **26**, 23 (1973).

157. A.M. Yaglom, *An Introduction to the Theory of Stationary Random Functions*, Prentice-Hall, Englewood Cliffs, N.J. (1962).

158. H. Cramér and M.R. Leadbetter, *Stationary and Related Stochastic Processes,* Wiley, New York (1967).

159. E.E. Slutsky, *Sur les fonctions aléatoires presque periodiques et sur la décomposition des fonctions aléatoires stationnaires en composantes,* Actual. Sci. Ind. No. 738, Herman et Cie, Paris (1933).

160. P.H. Diananda, Some Probability Limit Theorems with Statistical Applications, *Proc. Cambridge Philos. Soc.* **49**, 239 (1953).

161. U. Grenander, *Probabilities on Algebraic Structures,* Wiley, New York (1963).

162. L. Arnold, On the Asymptotic Distribution of the Eigenvalues of Random Matrices, *J. Math. Anal. App.* **20**, 262 (1967).

163. W.H. Olson and V.R.R. Uppuluri, in *Probability Theory*, Vol. 3 of Proceedings of the Sixth Berkeley Symposium on Mathematical Statistics and Probability, Ed. L.M. Le Cam, J. Neyman, and E.L. Scott, University of California, Berkeley (1972).

164. K.K. Mon and J.B. French, Statistical Properties of Many-Particle Spectra, *Ann. Phys.* (N.Y.) **95**, 90 (1975).

165. F.J. Dyson, Correlations between Eigenvalues of a Random Matrix, *Commun. Math. Phys.* **19**, 235 (1970).

166. M.L. Mehta, A Note on Correlations between Eigenvalues of a Random Matrix, *Commun. Math. Phys.* **20**, 245 (1971).

167. K. Ideno and M. Ohkubo, Nonrandom Distributions of Neutron Resonance Levels, *J. Phys. Soc. Japan* **30**, 620 (1971).

168. K. Ideno, Nonstatistical Behaviors of the Level Spacing Distributions in Neutron Resonances, *J. Phys. Soc. Japan* **37**, 581 (1974).

169. T.A. Brody and P.A. Mello, An Ergodic Property of Orthogonal Ensembles of Random Matrices, *Phys. Lett. A* 37, 429 (1971).

170. P.A. Mello and T.A. Brody, A Different Proof of the Maxwell-Boltzmann Distribution, *Amer. J. Phys.* **40**, 1239 (1972).

171. M.G. Kendall and A. Stuart, *The Advanced Theory of Statistics,* 2nd ed., Hafner, New York (1967).

172. R.J. Eden, *High Energy Collisions of Elementary Particles*, Cambridge University Press (1967).

173. A.O. Barut, *The Theory of the Scattering Matrix,* McMillan Co., New York (1967).

174. J. Richert and H.A. Weidenmüller, Equality of Energy Average and Ensemble Average in the Statistical Theory of Nuclear Reactions, *Phys. Rev. C* **16**, 1309 (1977).

175. J.B. French, P.A. Mello, and A. Pandey, Statistical Properties of Many-Particle Spectra, II. Two-Point Correlations and Fluctuations, *Ann. Phys.* (N.Y.) **113**, 277 (1978).

176. D. Agassi, H.A. Weidenmüller, and G. Mantzouranis, The Statistical Theory of Nuclear Reactions for Strongly Overlapping Resonances as a Theory of Transport Phenomena, *Physics Reports* 22, 145 (1975).

BIBLIOGRAPHY

S. Ayik and J.N. Ginocchio, Shell Model Level Densities for Light Nuclei, *Nucl. Phys. A* **234**, 13 (1974).

J. Barojas, E. Cota, E. Blaisten-Barojas, J. Flores, and P.A. Mello, Studies on the Problem of Small Metallic Particles. I. Spectrum Fluctuations in a Two-Dimensional Model and the Associated Specific Heat, *Ann. Phys.* (N.Y.) **107**, 95 (1977).

M.S. Bartlett, *An Introduction to Stochastic Processes*, Cambridge University Press, Cambridge, England (1966).

F. Becvar, R.E. Chrien, and O.A. Wasson, A Study of the Distribution of Partial Radiative Widths and Amplitudes for ^{149}Sm(n, γ)^{150}Sm, *Nucl. Phys. A* **236**, 198 (1974).

M. Beer, M.A. Lone, R.E. Chrien, O.A. Wasson, M.R. Bhat, and H.R. Muether, Correlations in Partial Widths of Neutron-Induced Reactions, *Phys. Rev. Lett.* **20**, 340 (1968).

J.M. Blatt and V.F. Weisskopf, *Theoretical Nuclear Physics*, Wiley, New York (1952).

C. Bloch, Statistical Theory of Nuclear Reactions as a Communication Problem. I. General Method. II. Constant and One-Level Resonating Amplitude, *Nucl. Phys. A* **112**, 257, 273 (1968).

O. Bohigas, J. Flores, J.B. French, M.J. Giannoni, P.A. Mello, and S.S.M. Wong, Recent Results on Energy-Level Fluctuations, *Phys. Rev. C* **10**, 1551 (1974).

O. Bohigas and M.J. Giannoni, Level Density Fluctuations and Random Matrix Theory, *Ann. Phys.* (N.Y.) **89**, 393 (1975).

L.M. Bollinger, in *Experimental Neutron Resonance Spectroscopy*, J.A. Harvey, Ed., Academic, New York (1970).

T.A. Brody, E. Cota, J. Flores, and P.A. Mello, Level Fluctuations: A General Property of Spectra, *Nucl. Phys. A* **259**, 87 (1976).

B.V. Bronk, Accuracy of the Semicircle Approximation for the Density of Eigenvalues of Random Matrices, *J. Math. Phys.* **5**, 215 (1964).

B.V. Bronk, Exponential Ensemble for Random Matrices, *J. Math. Phys.* **6**, 228 (1965).

H.S. Camarda, Upper Limit on a Time Reversal Noninvariant Part of Wigner's Random Matrix Model, *Phys. Rev. C* **13**, 2524 (1976).

H.S. Camarda, H.I. Liou, F. Rahn, G. Hacken, M. Slagowitz, W.W. Havens, Jr., J. Rainwater, and S. Wynchank, in *Statistical Properties of Nuclei*, J.B. Garg, Ed., Plenum, New York (1972).

H.S. Camarda, H.I. Liou, G. Hacken, F. Rahn, W. Makofske, M. Slagowitz, S. Wynchank, and J. Rainwater, Neutron Resonance Spectroscopy. XII. The Separated Isotopes of W, *Phys. Rev. C* 8, 1813 (1973).

F.S. Chang and J.B. French, Energy Dependence of Expectation Values in Many-Particle Spectroscopy, *Phys. Lett. B* **44**, 131 (1973).

F.S. Chang, J.B. French, and T.H. Thio, Distribution Methods for Nuclear Energies, Level Densities and Excitation Strengths, *Ann. Phys.* (N.Y.) **66**, 137 (1971).

F.S. Chang and A. Zuker, Validity of Spectra Distribution Methods in Spectroscopy, *Nucl. Phys. A* **198**, 417 (1972).

C.F. Clement, Theory of Overlap Functions. I. Single Particle Sum Rules and Centre-of-Mass Corrections, *Nucl. Phys. A* **213**, 469 (1973).

C.F. Clement and S.M. Perez, Non-Energy Weighted Sum Rule Analysis of One Nucleon Transfer Data on ^{41}Ca, ^{43}Ca, ^{45}Sc, ^{49}Ti, and ^{51}V and the Validity of the Shell Model, *Nucl. Phys. A* **284**, 469 (1977).

C. Coceva and M. Stefanon, Experimental Aspects of the Statistical Theory of Nuclear Spectra Fluctuations, *Nucl. Phys. A* **315**, 1 (1979).

H. Cramér, *Mathematical Methods of Statistics*, Princeton University Press, Princeton, N.J. (1946).

F. Cristofori, P.G. Sona, and F. Tonolini, The Statistics of the Eigenvalues of Random Matrices, *Nucl. Phys.* 78, 553 (1966).

B.J. Dalton, S.M. Grimes, J.P. Vary, and S.A. Williams, Eds., *Theory and Applications of Moment Methods in Many-Fermion Spaces*, Plenum, New York (1980).

R. Denton, B. Mühlschlegel and D.J. Scalapino, Thermodynamic Properties of Electrons in Small Metal Particles, *Phys. Rev. B* 7, 3589 (1973).

C. DeWitt and V. Gillet, Eds., *Physique Nuléaire,* Gordon and Breach, New York (1969).

J.P. Draayer, J.B. French, and S.S.M. Wong, Strength Distributions and Statistical Spectroscopy. I. General Theory. II. Shell-Model Comparisons, *Ann. Phys.* (N.Y.) **106**, 472, 503 (1977).

F.J. Dyson, A Class of Matrix Ensembles, *J. Math. Phys.* **13**, 90 (1972).

K.A. Eberhard, P. von Brentano, M. Böhning, and R.O. Stephen, An Explicit Expression for the Average Total Width of Compound Levels, *Nucl. Phys. A* **125**, 673 (1969).

S.F. Edwards and R.C. Jones, The Eigenvalue Spectrum of a Large Symmetric Random Matrix, *J. Phys. A* **9**, 1595 (1976).

S.F. Edwards and M. Warner, The Effect of Disorder on the Spectrum of a Hermitian Matrix, *J. Phys. A* **13**, 381 (1980).

C.A. Engelbrecht and H.A. Weidenmüller, Hauser-Feshbach Theory and Ericson Fluctuations in the Presence of Direct Reactions, *Phys. Rev. C* **8**, 859 (1973).

T. Ericson, A Theory of Fluctuations in Nuclear Cross Sections, *Ann. Phys.* (N.Y.) 23, 390 (1963).

H. Feshbach, A.K. Kerman, and S. Koonin, The Statistical Theory of Multi-Step Compound and Direct Reactions, *Ann. Phys.* (N.Y.) **125**, 429 (1980).

R. Fortet, in *Some Aspects of Analysis and Probability*, Surveys in Applied Mathematics, 4, Wiley, New York (1958).

J.B. French, in *Many-Body Description of Nuclei and Reactions*, International School of Physics, Enrico Fermi, Course 36, C. Bloch, Ed., Academic, New York (1966).

J.B. French and J.P. Draayer, in *Group Theoretical Methods in Physics,* W. Beiglböck, A. Böhm, and E. Takasugi, Eds., Springer, Berlin (1979).

J.B. French, P.A. Mello, and A. Pandey, Ergodic Behavior in the Statistical Theory of Nuclear Reactions, *Phys. Lett. B* **80**, 17 (1978).

J.B. French and S.S.M. Wong, Some Random-Matrix Level and Spacing Distributions for Fixed-Particle-Rank Interactions, *Phys. Lett. B* **35**, 5 (1971).

J.D. Garrison, Statistical Analysis of Nuetron Resonance Parameters, *Ann. Phys.* (N.Y.) **30**, 269 (1964).

A. Gervois, Level Densities for Random One- or Two-Body Potentials, *Nucl. Phys. A* **184**, 507 (1972).

G. Hacken, H.I. Liou, H.S. Camarda, W.J. Makofske, F. Rahn, J. Rainwater, M. Slagowitz, and S. Wynchank, Neutron Resonance Spectroscopy. XVI. ^{113}In, ^{115}In, *Phys. Rev. C* **10**, 1910 (1974).

M. Handelman, A Generalized Eigenvalue Distribution, *J. Math. Phys.* **19**, 2509 (1978).

E.J. Hannan, *Time Series Analysis*, Methuen, New York (1960).

W. Hauser and H. Feshbach, The Inelastic Scattering of Neutrons, *Phys. Rev.* **87**, 366 (1952).

H.M. Hofmann, J. Richert, and J.W. Tepel, Direct Reactions and Hauser-Feshbach Theory II. Numerical Methods and Statistical Tests, *Ann. Phys.* (N.Y.) **90**, 391 (1975).

H.M. Hofmann, J. Richert, J.W. Tepel, and H.A. Weidenmüller, Direct Reactions and Hauser-Feshbach Theory, *Ann. Phys.* (N.Y.) **90**, 403 (1975).

A.P. Jain and J. Blons, Higher-Order Spacing Correlations in ^{239}Pu and ^{240}Pu, *Nucl. Phys. A* **242**, 45 (1975).

R.C. Jones, J.M. Kosterlitz, and D.J. Thouless, The Eigenvalue Spectrum of a Large Symmetric Random Matrix with a Finite Mean, *J. Phys. A* **11**, L45 (1978).

M. Kawai, A.K. Kerman, and K.W. McVoy, Modification of Hauser-Feshbach Calculations by Direct-Reaction Channel Coupling, *Ann. Phys.* (N.Y.) **75**, 156 (1973).

M.G. Kendall and A. Stuart, *The Advanced Theory of Statistics*, 3rd ed., Vol. 1, Hafner, New York (1969).

A.K. Kerman and A.F.R. de Toledo Piza, Studies in Isobaric Analog Resonances. II. Fine Structure, *Ann. Phys.* (N.Y.) **48**, 173 (1968).

A.K. Kerman and A. Sevgen, On the Role of Unitarity in Statistical Theories of Nuclear Reactions, *Ann. Phys.* (N.Y.) **102**, 570 (1976).

G.J. Kirouac and H.M. Eiland, Total Neutron Cross Section for ^{151}Sm and Nuclear Systematics of the Sm Isotopes, *Phys. Rev. C* **11**, 895 (1975).

T.J. Krieger, Statistical Theory of Nuclear Cross Section Fluctuations, *Ann. Phys.* (N.Y.) **42**, 375 (1967).

A.M. Lane and J.E. Lynn, Theory of Radiative Capture in the Resonance Region, *Nucl. Phys.* **17**, 563 (1960).

A.M. Lane, J.E. Lynn, and J.D. Moses, Line Shape in Weak and Intermediate Coupling: Theory and Practical Fitting Procedures, *Nucl. Phys. A* **232**, 189 (1974).

A.M. Lane, J.E. Lynn, and J.D. Moses, Comment on "Statistical Significance of Spreading Widths for Doorway States," *Phys. Rev. C* **20**, 2435 (1979).

A.M. Lane and R.G. Thomas, R-Matrix Theory of Nuclear Reactions, *Rev. Mod. Phys.* **30**, 257 (1958).

H.S. Leff, Asymptotic Densities in Statistical Ensembles, *Phys. Rev.* **136**, 355A (1964).

H.I. Liou, H.S. Camarda, and F. Rahn, Application of Statistical Tests for Single-Level Populations to Neutron-Resonance-Spectroscopy Data, *Phys. Rev. C* **5**, 1002 (1972).

H.I. Liou, H.S. Camarda, S. Wynchank, M. Slagowitz, G. Hacken, F. Rahn, and J. Rainwater, Neutron-Resonance Spectroscopy. VIII. The Separate Isotopes of Erbium: Evidence for Dyson's Theory Concerning Level Spacings, *Phys. Rev. C* **5**, 974 (1972).

H.I. Liou, G. Hacken, J. Rainwater, and U.N. Singh, Neutron Resonance Spectroscopy: The Separated Isotopes of Dy, *Phys. Rev. C* **11**, 462, (1975).

J.E. Lynn, *The Theory of Neutron Resonance Reactions*, Clarendon, Oxford (1968).

W.M. MacDonald, Statistical Significance of Spreading Widths for Doorway States, *Phys. Rev. C* **20**, 426 (1979).

W.M. MacDonald, Reply to "Comment on 'Statistical Significance of Spreading Widths for Doorway States'," *Phys. Rev. C* **21**, 1642 (1980).

J.F. McDonald, Some Mathematically Simple Ensembles of Random Matrices which Represent Hamiltonians with a Small Time-Reversal-Noninvariant Part, *Nuovo Cimento B* **57**, 95 (1980).

K.W. McVoy, L. Heller, and M. Bolsterli, Optical Analysis of Potential Well Resonances, *Rev. Mod. Phys.* **39**, 245 (1967).

K.W. McVoy and P.A. Mello, Strong Absorption and the Distribution of Zeros in the *S*-Matrix, *Nucl. Phys. A* **315**, 391 (1979).

M.L. Mehta and N. Rosenzweig, Distribution Laws for the Roots of a Random Antisymmetric Hermitian Matrix, *Nucl. Phys. A* **109**, 449 (1968).

P.A. Mello, J. Flores, T.A. Brody, J.B. French, and S.S.M. Wong, in *Proceedings of the International Conference on Interactions of Neutrons with Nuclei*, E. Sheldon, Ed., Nat. Tech. Information Service, Springfield, Virginia (1976).

P.A. Mello and T.H. Seligman, On the Entropy Approach to Statistical Nuclear Reactions, *Nucl. Phys. A* **344**, 489 (1980).

G.E. Mitchell, in *Theory and Applications of Moment Methods in Many-Fermion Spaces,* B.J. Dalton *et al.*, Eds., Plenum, New York (1980).

P.A. Moldauer, Statistical Theory of Nuclear Collision Cross Sections, *Phys. Rev.* **135**, 642B (1964).

P.A. Moldauer, Statistical Theory of Nuclear Collision Cross Sections. II. Distributions of the Poles and Residues of the Collision Matrix, *Phys. Rev.* **136**, 947B (1964).

P.A. Moldauer, Averaging Methods in Nuclear Reaction Theory, *Phys. Rev. Lett.* 23, 708 (1969).

P.A. Moldauer, Why the Hauser-Feshbach Formula Works, *Phys. Rev. C* **11**, 426 (1975).

J.E. Monahan and N. Rosenzweig, Analysis of the Distribution of the Spacings Between Nuclear Energy Levels. II., *Phys. Rev. C* **5**, 1078 (1972).

W. Nörenberg and H.A. Weidenmüller, *Introduction to the Theory of Heavy-Ion Collisions*, Lecture Notes in Physics 51, J. Ehlers, K. Hepp, R. Kippenhahn, H.A. Weidenmüller, and J. Zittartz, Eds., Springer, Berlin (1976).

C.F. Perdrisat, Survey of Some Systematic Properties of the Nuclear *E*1 Transition Probability, *Rev. Mod. Phys.* 38, 41 (1966).

Yu. V. Prohorov and Yu. A. Rozanov, *Probability Theory*, Die Grundlehren der Mathematischen Wissenschaften in Einzeldarstellunger, Vol. 157, Springer, Heidelberg (1969).

K.F. Ratcliff, Applications of Spectral Distributions in Nuclear Spectroscopy, *Phys. Rev. C* 3, 117 (1971).

J. Riordan, *Combinatorial Identities,* Wiley, New York (1968).

N. Rosenzweig and C.E. Porter, "Repulsion of Energy Levels" in Complex Atomic Spectra, *Phys. Rev.* **120**, 1698 (1960).

N. Rosenzweig, J.E. Monahan, and M.L. Mehta, Perturbation of the Statistical Properties of Nuclear States and Transitions by Interactions That Are Odd Under Time Reversal, *Nucl. Phys. A* **109**, 437 (1968).

E. Sheldon, Ed., *Proceedings of the International Conference on the Interaction of Neutrons with Nuclei*, Lowell, Massachusetts, Tech. Information Center, ERDA, Washington, D.C. (1976).

J. Touchard, R.U. Haq, and R. Arvieu, An Ensemble of Random Particle-Hole Matrices with Collective Eigenstates, *Z. Physik A* **282**, 191 (1977).

N. Ullah, Invariance Hypothesis and Higher Correlations of Hamiltonian Matrix Elements, *Nucl. Phys.* **58**, 65 (1964).

N. Ullah, in *International Nuclear Physics Conference*, R.L. Becker, Ed., Academic, New York (1967).

O.A. Wasson, R.E. Chrien, G.G. Slaughter, and J.S. Harvey, Distribution of Partial Radiation Widths in $^{238}U(n, \gamma)^{239}U$, *Phys. Rev. C* **4**, 900 (1971).

H.A. Weidenmüller, Level-Level Correlations in Hauser-Feshbach Theory and Moldauer's Sum Role for Resonance Reactions, *Phys. Rev. C* **9**, 1202 (1974).

E.P. Wigner, On the Statistical Distribution of the Widths and Spacings of Nuclear Resonance Levels, *Proc. Cambridge Philos. Soc.* **47**, 790 (1951).

E.P. Wigner, On the Connection Between the Distribution of Poles and Residues for an *R* Function and Its Invariant Derivative, *Ann. Math.* **55**, 7 (1952).

E.P. Wigner, in *International Conference on the Neutron Interactions with the Nucleus*, Columbia University, 1957, Columbia University Report No. CU-175 (TID-7547, Columbia University) (1957).

W.M. Wilson, E.G. Bilpuch, and G.E. Mitchell, Applications of Statistical Tests to Proton Resonances in ^{45}Sc and ^{49}V, *Nucl. Phys. A* **245**, 285 (1975).

L. Wolfenstein, Conservation of Angular Momentum in the Statistical Theory of Nuclear Reactions, *Phys. Rev.* 82, 690 (1951).

S.S.M. Wong and J.B. French, Level-Density Fluctuations and Two-Body *versus* Multi-Body Interactions, *Nucl. Phys. A* **198**, 188 (1972).

S.S.M. Wong and G.D. Lougheed, A Study of ^{28}Si Using Shell-Model and Statistical Spectroscopic Methods, *Nucl. Phys. A* **295**, 289 (1978).

E. Yépez, Doctoral Dissertation, Universidad Nacional Autónoma de México (1975).

INDEX

Since the Bibliography (given on pages 169-178) is alphabetically arranged, references cited in it are not included in this Index.

A

Absence of degeneracy, 47
Absence of small spacings, 47
Actual density of levels, 54
Additive statistic, 137
Adjacent roots, interval between, 46
Admissible Hamiltonians, 4
Agassi, D., 150, 167
Ajzenberg, F., 4, 5, 152
Al-Ani, S., 102, 160
Algebra, quaternion, 69, 70-72
Algebra of symplectic group, 67
Algebraic equation, 25
Alternate levels, 50
Alternative procedures, 123-127
Andreief, C., 90, 160
Angles,
  consecutive, 24
  functions of energy levels, 24
  range of, 24
Angles of rotation, 77
Angular momenta, 4, 5
  orbital part of, 18
Angular momentum, 47
  conservation of, 16
  total, 17, 18, 19, 21
Angular momentum quantum number, 5, 7
Anomalous functions, 135
Antiinvariant operator, 58
Antilinear operator, 16, 17
Antiunitary operator, 17
Applications of multivariate distributions, 115-127
Approximation,
  continuous, 54
  smoothed, 54
Arbitrary symplectic matrix, 72
Arnold, L., 136, 166
Aspects, random, 51
Atom, 2, 19, 29
  hydrogen, 9, 10
Atomic physics, 30, 59
Atomic surface, 57
Atomic system, 2
Autocovariance function, 133, 135, 139, 141, 145
Automorphism, 28, 30, 73
Average,
  ensemble, 146, 147
  running, 54
  spectral, 135, 141, 143, 146, 147
Average level spacing, 29
Average spacing, 33, 37, 41, 46, 54
Average widths, 7
Averages,
  rms, 51

[Averages,]
time, 3, 41
Averaging,
phase and time, 42
spectral, 130, 131, 132, 133, 135, 136, 137, 142
Averaging interval, 132, 133, 144
Averaging process, 4
Axes, spatial, 16
Axis,
energy, 42, 131
time, 42

B

Balescu, R., 130, 165
Balian, R., 86, 158
Band, rotational, 5
Banded diagonal matrix, 19
Baranger, M., 2, 151
Bargmann, R., 118, 163
Baronk, B. V., 26, 154
Bartlett, M. S., 118, 164
Barut, A. O., 150, 167
Basic space, 28
Basic statistical hypothesis, 24
Basis, 19
unitary, 19
Basis vector, 146
Behavior,
ergodic, 62, 129, 133, 134, 135, 136, 142, 146, 149
secular, 50, 56
Behavior of strength distribution, 146-150
Beryllium, 4
energy levels of, 4
Binary fluctuations, 134
Birkhoff, G. D., 4, 152
Black box, 3, 28
Bohigas, O., 85, 158
Bonferroni's inequality, 121
Boron, 4
energy levels of, 4
Boundary condition, 10
Boundary, nuclear, 7
Box, G. E. P., 113, 114, 162
Box's method, 114
Brillinger, D. R., 88, 116, 159
Bremmer, H., 10, 153
Brody, T. A., 30, 50, 60, 129, 146, 155, 166
Brody *et al.*, 31, 33, 35, 37, 41, 51, 55, 61, 62, 87, 131, 136, 139, 146, 150
Brownian motion, 50, 86
Brownian motion model, 50
Brownian particles, 86

C

Calculation, shell-model, 33, 38, 39, 51, 60, 64, 65
Calculations,
computer, 42
Monte-Carolo, 49, 65
Canonical correlation matrix, 126
Canonical correlations, 125
Canonical ensembles, 136
Canonical transformation, 16, 19
Canonical transformation group, 18, 21
Canonical unitary transformation, 20
Carbon, energy levels of, 4
Carmeli, M., 77, 87, 153, 157
Carter, R. S., 157
Central chi-square variates, 109, 114
Central complex bivariate beta matrix, 107
Central complex bivariate F matrix, 107
Central complex Gaussian matrix, 96
Central complex multivariate beta matrix, 96, 103, 109, 110
Central complex multivariate F matrix, 96, 120
Central complex Wishart distribution, 96, 97
Central complex Wishart matrix, 96, 99, 102, 103, 109, 116, 119, 123
Central limit theorem, 96
Central multivariate beta matrix, 99
Central Wishart matrix, 102
Centroid, 32, 60, 61, 140, 146, 149
Chang, F. S., 85, 158
Chang, T. C., 86, 98, 103, 110, 114, 158, 160, 162, 163
Characteristic function, 4, 5, 58, 100
Characteristic value, 2, 4, 59
Characteristic values, 86

Characteristic value of complex
Hermitian matrix,46
Characteristic values of matrix, 46
Characteristic values of real symmetric
matrix, 46
Characteristic vector, 2
Chattopadhyay, A. K., 86, 158
Chevalley, C., 20, 67, 154
Christenson, J. H., 7, 153
Circle, unit, 24, 26, 70, 72
Circular ensembles, 141
Circular orthogonal ensemble, 49, 50
Circular symplectic ensemble, 50
Circular unitary ensemble, 50
Class of real symmetric operators, 7
Class of systems, wide, 54
Classical mechanics, 3
Classical polynomials, 50
Classical statistical mechanics, 33, 41
Cluster functions, 142, 145
Coeseter, F., 28, 150, 154
Coherence,
multiple, 99
partial, 99
Collective behavior and strength
fluctuations, 60-62
Collective nuclei, 41
Collectivity, 61
Combination of spin, invariant, 15
Commutation relation, 17
Commuting Hermitian quaternion-real
matrices, 70
Commuting operators, 18
Compact space, 26
Comparison with statistical mechanics,
3-4
Complementary set, 92, 93
Complete spectrum, 132
Completely random sequence levels,
38
Complex canonical correlation matrix,
103
Complex conjugation, 17
Complex conjugation operator, 17
Complex Gaussian ensemble matrix,
103
Complex Gaussian stochastic process,
126
Complex Hermitian matrix, character-
istic values of, 46
Complex matrices, 20, 86
Complex multivariate beta matrix, 103
Complex multivariate distributions, 8,
87, 115
Complex multivariate normal distribu-
tion, 96, 110, 125
Complex multivariate normal popula-
tion, 110, 114
Complex multivariate processes, 126
Complex nuclei, 42
Complex nucleus, 2, 3
Complex numbers, 24
Complex quaternion, 68
Complex random matrices, 88
Complex spectra, 68
Complex system, internal energy of,
16
Complex systems, 1-2, 8, 9, 16, 41
Complex Wishart matrix, 87, 88, 96,
97, 99, 103, 116, 118, 119, 123,
125
Complex Wishart process, 127
Components,
eigenvector, 146
statistically independent, 24
Components of matrix elements, 85
Components of wavelengths, 54
Computer calculations, 42
Concepts, statistical, 11
Condition, normalization, 53, 79
Conditional probability, 31
Conditions,
boundary, 10
initial, 4
Confidence limits, 135
Conjecture, Wigner's, 47-49
Conjugate quaternion, 68
Conjugation, complex, 17
Consecutive angles, 24
Consecutive levels of actual system, 24
Conservation,
energy, 15
linear momentum, 16
Conservation of angular momentum,
16
Constant, Planck's, 5
Constant of motion, 4, 10, 16
Continuous approximation, 54
Continuous curves, 38
Convolution, 107
Continuous functions, time averages
of, 3

Continuum for positive energies, 10
Convex distribution, 26
Coordinate representation, 17, 18
Coordinate system, 3, 16, 58
Coordinate vectors, interparticle, 15
Coordinates, 3, 4, 16
  generalized, 3
Correlation functions, 139
Covar (covariance of two random variables), 147
Covariance, 148
Covariance matrix, 95, 99, 110, 114, 115, 118, 122, 125
CP invariance, 7
Cramer, H., 130, 165
Cronin, J. W., 7, 153
Cross sections, fluctuations in, 83
Crystalline electric field, 19
Curves, continuous, 38
Cut-off parameter, 56

D

Das, M. M., 114, 163
Data,
  experimental, 47, 133
  nuclear-table, 56
  slow-neutron, 56
Decay law, 12
Decay of compound nucleus, 83
Decaying radiative source, 11, 12
Definite masses, 3
Degeneracy, 10
  absence of, 47
  doublet, 19
  Kramers, 19-20, 70
Degree of irregularity, 2
Delta function, 51, 135, 142, 146
Delta function singularity, 135, 140, 141
Densities,
  joint-probability, 140
  level, 83
Density, 132
  Gaussian, 56, 131, 149
  joint, 97, 98, 99, 100, 102, 103, 108, 109, 118
  level, 45, 49, 50, 86, 135, 136
  mean, 11
[Density]
  mean level, 49
  probability, 32, 53
Density function, 96
Density in neighborhood of lowest state, 26
Density in range of semicircle law, 26
Density matrix, 87, 115, 125
Density of discrete spectrum, 51
Density of energy levels, 7, 26
Density of levels, 7, 26, 32, 54
  actual, 54
Density of prime numbers, 10, 11
Density of roots, joint, 97, 100, 103, 108, 109
Desjardins, J. S., 2, 152
Determinant, 78
  distribution of, 109, 110
Deviation,
  level, 54
  level-to-level, 56
  rms, 51, 56
Deviation of spectrum, 51
Diagonal matrix, 77
  banded, 19
Diananda, P. H., 135, 165
Diananda's theorem, 135
Different effective interactions, 56
Differential probability, 12, 77
Dimension, infinite-, 3
Dimension of random matrix, 26
Dimension of space, 76
Dimensionality, 56
  high, 3
Dimensionality of space, 53
Dimensions, odd number of, 16
Dipole, magnetic, 12
Dipole moment, 58
Direction of time, 16
Discrete energy levels, 42
Discrete levels, 12, 42
Discrete nuclear spectra, 41
Discrete variables, 31
Discrimination between multivariate normal population, 122-123
Distance, mean, 12
Distribution, 25, 26, 39, 59, 62, 96, 106
  behavior of strength, 146-150
  central complex Wishart, 96, 97

[Distribution]
complex multivariate normal, 96, 110
convex, 26
eigenvalue, 44, 48, 50, 75
eigenvalue-eigenvector, 8, 42
eigenvector, 44, 75
finite gap in, 86
Gaussian, 26, 56, 58, 85
joint, 118
level, 25
level-spacing, 60
limiting, 54
matrix element, 43, 44
nearest-neighbor spacing, 12, 31, 38, 45, 46, 47, 50, 56
orthogonal, 26
Pearson's, 108, 114
Poisson, 12, 32, 33, 38, 39, 41, 47, 142
Porter-Thomas, 58-60, 62, 64, 65, 146, 147, 149
Rayleigh, 32
semicircle, 25-26, 136
Semiellipse, 25, 26
single-eigenvalue, 45, 48
spacing, 38, 41, 46, 47, 50
strength, 146
tables for, 108
uniform, 26
uniform probability, 24
widths, 7-8, 57-66
Wigner's, 32, 33, 38, 39, 41
Wigner's semicircle, 136
Wishart, 85
Distribution function, 23, 24, 44, 47, 53, 54, 64, 76-78, 113, 114, 131
expansion of, 54
specification of, 79-80
Wigner's, 47
Distribution of determinant, 109, 110
Distribution of eigenvalues, 44, 48, 50, 75, 123, 124
Distribution of likelihood ratio test statistic, 110
Distribution of nearest-neighbor spacings, 42
Distribution of ratio of extreme roots, 109
Distribution of roots, 86, 88, 95, 102 103, 109
Distribution of spacings, 87
Distribution of widths, 7-8, 57-66
Distributions, 87-88
applications of multivariate, 115-127
complex multivariate, 8, 87, 115
eigenvalue, 44, 48, 50
eigenvector, 44
energy level, 25
higher-order spacing, 49
joint, 109
level, 24
marginal, 88, 109
multivariate, 8, 87-114
non-random, 143
rate of change with dimension of, 49
spacing, 45, 49, 50
strength, 129
traces of complex random matrices, 88
Distributions of likelihood ratio test statistics, 109-114
Distributions of matrix elements, 136
Distributions of ratios of roots, 108-109
Distributions of traces, 88, 107
Distributions of some random matrices, 94-100
Doorway states, 62
Double root of matrix, 46
Doublet degeneracy, 19
Doubly degenerate energy levels, 70
Doubly magic nuclei, 41
Doubly parity degenerate levels, 10
Dresner, L., 50, 156
Dubman, M. R., 127, 164
Dyson, F. J., 2, 19, 21, 24, 29, 47, 49, 50, 51, 67, 135, 138, 140, 151, 154, 155, 156, 166
Dyson's *k*-level cluster functions, 140, 142
Dyson-Mehta formalism, 51

E

Eden, R. J., 150, 166
Effective interactions, 56
Eigenfunctions, 2, 146
Eigenstates, 16, 17

Eigenvalue distributions, 44, 48, 50, 75, 123, 124
Eigenvalue variables, 78
Eigenvalue-eigenvector distributions, 8, 42
Eigenvalue-eigenvector distributions of Gaussian ensemble, 42, 43-56
Eigenvalues, 2, 12, 24, 54, 70, 77, 78, 79, 117, 123, 125, 126, 150
Eigenvector components, 146
Eigenvector distributions, 44, 75
Eigenvector variables, 78
Eigenvectors, 45, 77, 78, 79, 146, 150
Electric field, crystalline, 19
Electric quadrupole, 12
Electromagnetic interaction, 16
Element,
  line, 76
  volume, 76, 77
Element of partition, 106
Elementary symmetric function, 104, 114, 118
Elements,
  matrix, 58, 59, 79, 86
  quaternion, 69
  statistics of matrix, 58
Emission, 12
  neutron, 59
  particle, 12
Energies, continuum for positive, 10
Energy, 4, 7, 15, 16, 19, 26, 42, 84
  exponential increase with, 26
  kinetic, 16
  total, 25
Energy axis, 42, 131
Energy conservation, 15
Energy level, 31
  lowest, 7, 26
Energy level distributions, 25
Energy level fluctuations, 60
Energy level sequences, 50
Energy level spacings, 10, 44
Energy levels, 1, 2, 3, 4, 7, 10, 33, 70, 87, 137, 145, 150
  angles functions of, 24
  beryllium, 4
  boron, 4
  caron, 4
  density of, 7, 26
  distribution of, 25
[Energy levels]
  doubly degenerate, 70
  doubly parity degenerate, 10
  ensemble of, 3
  exact values of, 4
  functions of, 24
  ^{180}Hf, 5
  low-lying, 7
  nucleus with an infinite number of, 24
  parity degenerate, 10
  positions of, 10
  repulsion of, 46
  sequence of, 50
  statistical theory of, 42, 81
  structure of, 3
  theoretical calculation of, 13
  ^{239}U, 7
  widths of, 12, 57-58
Energy levels and random matrices, 83-84
Energy levels spectra, 38
Energy of excitation, 25
Energy of physical system, 15
Energy operators, 7
Energy ordering, 147
Energy range, 25
Ensemble, 2, 24, 25, 26, 27, 30, 33, 42, 75, 129, 130, 132
  admissible Hamiltonians, 4
  circular orthogonal, 49, 50
  circular symplectic, 50
  circular unitary, 50
  complex Hermitian matrices, 59
  energy levels, 3
  Gaussian, 8, 21, 23-24, 25, 28, 29, 42, 43-45, 47, 49, 50, 73, 75-81, 143, 146, 150
  Gaussian orthogonal, 49, 136, 138
  Hamiltonians, 33, 150
  Hamiltonians of the, 59
  Hermitian matrices, 3
  invariance of, 28
  orthogonal, 8, 21, 23, 24-25, 26, 29, 50, 72, 75, 80, 81, 142
  Poisson, 141, 146
  S-matrix, 130, 150
  states, 3
  symplectic, 29, 50, 72-73, 80, 81, 142

[Ensemble]
systems, 29, 50, 72-73, 80, 81, 142
trangent, 131
uniform, 30
unitary, 8, 28, 29-42, 50, 80-81, 142
Wishart, 26-27, 86
Ensemble average, 146, 147
Ensemble average of function, 131
Ensemble-averaged local spacing, 130
Ensemble averages, 129
Ensemble of Hamiltonians, 150
Ensemble variances, 143
Ensembles, 8, 41, 130
canonical, 136
circular, 141
Gaussian, 81, 136, 137, 138, 139, 141, 145
Gaussian and orthogonal, 23-28
generalized, 50
Poisson, 136, 139
random matrix, 130
Ensembles of statistical mechanics, 3
Entire spectrum, 39, 51
Entropy theorem, 84
Equal pairs, 70
Equal probability, 27
Equal probability and invariance, 27-28
Equally probable interactions, 24
Equally probable systems, 28
Equation,
algebraic, 25
real roots of, 25
Schrödinger, 1, 2, 11, 12, 17, 19, 77
Equations of motion, 84
integration of, 41
Equidistant levels, 7
Equipartition theorem, 84
Ergodic behavior, 62, 129, 133, 134, 135, 136, 142, 146, 149
Ergodic measures, 135
Ergodic properties of random matrices, 8, 42, 129-150
Ergodic stationary random process, 130
Ergodicity, 42, 129, 130, 132, 134, 135, 136, 139, 141, 146, 147, 148, 149, 150
locally generated, 130
strength, 146
Ergodicity in reactions, 130
Ergodicity of fluctuation measures, 140
Error,
fractional, 137
mean-square, 142
relative, 135, 145
statistical, 135
Error function, 48
Essence of ergodic, 42
Estimate, Poisson, 142, 143
Estimator, 137, 138, 149
eTr (exponential of trace), 95
Even-even nuclei, 41
Even function, 133
Even parity, levels of, 10
Even solutions, 10
Even-spin Hamiltonian, 72
Exact percentage points, 102
Exact values of energy levels, 4
Examples, 4-7, 45
Excitation, total energy of, 25
Excitations, 54, 56
Excited states, 2, 10
Excited states, low, 13
Excited systems, 9-13
Excluded nuclei, 42
Expansion of distribution function, 54
Expectation values, 44, 57, 59, 75
Expectation values of operators, 12
Experimental data, 47, 133
Experimental data and Porter-Thomas distribution, 62-66
Experimenter, 124, 126
External crystalline electric field, 19
External field,
electric, 21
nonisotropic, 21
Extreme roots, 102

F

Factor, normalization, 45
Favro, L. D., 50, 156
Field,
crystalline electric, 19
external electric, 21
magnetic, 29
nonisotropic external, 21
zero, 30

Figure of merit, 135, 137, 143, 144
Final remarks, 150
Final state, 28
Finite and infinite matrices, 86
Finite gap in distribution, 86
First-excited state, 41
Fitch, V. L., 7, 153
Fixed interval, 138
Flores, J., 30, 50, 60, 85, 129, 155, 158
Fluctuation-free spectrum, 51
Fluctuation measures, 129, 130, 139, 142
  ergodicity of, 140
Fluctuation pattern, 39
Fluctuation properties of widths of energy levels, 57
Fluctuations, 42, 51, 54, 56, 61, 134
  binary, 134
  energy-level, 60
  ground-state, 41, 61
  residual level-to-level, 56
  separation of, 50
  strength, 60
Fluctuations about mean density, 11
Fluctuations in cross sections, 83
Fluctuations for pure sequences, 56
Formalism of Dyson and Mehta, 51
Formula,
  Poisson, 143
  Wigner's, 47
Fox, D., 50, 156
Fraction of levels, 7
Fractional error, 137
French, J. B., 30, 50, 60, 85, 129, 130, 136, 146, 150, 155, 158, 165, 166, 167
Function,
  autocovariance, 133, 135, 139, 141, 145
  characteristics, 4, 5, 58, 100
  delta, 51, 135, 141, 142, 145
  density, 96
  distribution, 23, 24, 44, 47, 53, 54, 64, 76-78, 113, 114, 131
  elementary symmetric, 104, 114, 118
  ensemble average of, 131
  error, 48
  even, 133
[Function]
  expansion of distribution, 54
  hypergeometric, 94, 95
  invariant, 24, 79
  joint characteristic, 99, 100
  k-point, 142, 145
  kernel, 50
  moment of, 104, 111
  one-point, 134
  spectral density, 149
  spectrally-smoothed density, 149
  square of wave, 7, 57
  staircase, 53
  state, 1
  symmetric, 89, 90, 91, 104
  time average of, 3, 41
  two-point, 139, 143, 149
  two-state, 16
  wave, 7, 12, 57
  weighting, 33
  Wigner's distribution, 47
Function of Hamiltonian matrix, 24
Functions
  anomalous, 135
  correlation, 139, 140
  Dyson's *k*-level cluster, 140, 142, 145
  higher-order, 139-146
  higher-order correlation, 130
  moments of elementary symmetric, 104-108
  strength, 83
Functions of energy levels, 24
Functions of time, 3, 4

G

Gamma-ray, 64, 66
Gap, finite, 86
Gaudin, M., 47, 155
Gaussian and orthogonal ensembles, 23-28
Gaussian density, 56, 131, 149
Gaussian distribution, 26, 56, 58, 85
Gaussian ensemble, 8, 21, 23-24, 25, 28, 29, 43-45, 47, 49, 50, 73, 75-81, 131, 143, 146, 150
  eigenvalue-eigenvector distribution of, 8, 42, 43-56

[Gaussian ensemble]
Wigner's asymptotic semicircle law for, 49
Gaussian ensembles, 81, 136, 137, 138, 139, 141, 145
Gaussian matrix, 98
Gaussian multiple time series, 87
Gaussian orthogonal ensemble, 49, 136, 138
Gaussian random variables, 23, 32, 62, 135
Gaussian stationary multiple time series, 118
Generalized coordinates, 3
Generalized ensembles, 50
Generalized momenta, 3
Giri, N., 110, 162
Global spectral averaging, 132
Global stationarity, 141
Goldstein, H., 4, 152
Good quantum numbers, 41
Goodman, N. R., 88, 96, 99, 109, 116, 127, 159, 162, 164
Grag, J. B., 44, 83, 155, 157
Gravitational interaction, 16
Grenander, U., 135, 136, 166
Ground state, 10, 39, 41, 42, 51, 60, 66
Ground-state fluctuations, 41, 61
Group, 27, 72, 77
algebra of symplectic, 67
canonical transformation, 18, 21
orthogonal, 19, 21, 78
quaternions and symplectic, 67-73
symplectic, 8, 20, 21, 66, 67
unitary, 18, 21, 30
Group measure, invariant, 30
Gunson, J., 50, 156
Gupta, A. K., 110, 162
Gurevich, I., 50, 156

H

h (Planck's constant), 5
Hafnium, 5, 6
Half-integer spin particles, 28
Half-integral spin, 19
Half-integral total angular momentum, 19, 21
Hamiltonian, 3, 4, 11, 15, 16, 19, 21, 24, 25, 33, 39, 58, 60, 61, 62, 65, 70, 76, 79, 84, 129
even spin, 72, 75, 76
Hermitian property of, 18
invariance of, 15, 21
nuclear, 84
odd spin, 70
proper, 4
real, 19, 23
real symmetric matrix, 23, 76
reality of, 24
realistic, 65
role of, 84
symmetric, 19, 23
time-reversal invariance of, 21
time-reversal invariant, 18
Hamiltonian matrix, 18-19, 24, 25, 39
Hamiltonian matrix element variables, 78
Hamiltonian of ensemble, 59
Hamiltonian of system, 75
Hamiltonian of system without time-reversal invariance, 30
Hamiltonian operator, 1, 3, 12
Hamiltonian spectrum, 149
Hamiltonian submatrix, 43
Hamiltonians,
ensemble of, 33
ensemble of admissible, 4
random, 86
Hannan, E. J, 88, 158
Harmonic oscillator, 9
Harvey, J. A., 6, 7, 57, 152, 157
Havens, W. W. Jr., 2, 44, 152, 155
Heavy nuclei, 2
Heavy nucleus, 51
Hermitian conjugate quaternion, 69
Hermitian conjugation, 69
Hermitian matrices, ensemble of, 3, 59
Hermitian matrix, 3, 30, 77, 94, 127
infinitesimal, 30
zonal polynomial of, 94
Hermitian operator, 1, 2, 7, 58
Hermitian property of Hamiltonian, 18
Hermitian quaternion-real matrix, 70
Hermitian random matrix, 96
Herz, C. S., 95, 160
High dimensionality, 3

High dimensionality space, 59
High excitation energy levels, 39, 87
Higher-energy region, 4
Higher-order correlation functions, 130
Higher-order functions, 139-146
Higher-order spacing distributions, 49
Higher-than-two-point measure, 145
Highly excited complex systems, 8, 9
Highly excited states, 2, 10
Highly excited systems, 9-13
Hilbert space, 3, 59, 84
Histogram, 25, 38, 39, 41, 62, 65, 149
Histogram of spectrum, 56
Histograms of nearest-neighbor spacings, 38
Homogeneity of space-time, 21
Hua, L. K., 19, 77, 95, 154, 157, 160
Hughes, D. J., 6, 7, 57, 152, 157
Hydrogen atom, 9, 10
Hydrogen *s* state, 10
Hypergeometric function, 94, 95
  approximation to, 95
Hypothesis, 110, 112, 113, 122, 123-127
  nested, 123
  testing, 116-122, 123-127

I

Ideal gas, 146
Ideno, K., 143, 166
Identifiable quantum numbers, 47
Identity operator, 17
Ignorance, 28
Improper transformation, 16
Independent particle model, 85, 86
Individual levels, 2
Individual spectrum, 42
Individual states, 38
Individual system, 41
Inference, 125
Infinite and finite matrices, 86
Infinite-dimension Hilbert Space, 3
Infinite-dimensional case, 50
Infinite-dimensional results, 47
Infinite matrices, 3
Infinite range, 24
Infinite square well, 9
Infinite-time limit, 132
Infinitesimal Hermitian matrix, 30
Infinitesimal interval, 31
Infinitesimal interval of length, 31
Infinitesimal neighborhood, 72
Information theory, 86
Initial conditions, 4, 84
Initial state, 28
Initial values, 4
Integer, 2, 11
Integral, probability, 88, 100, 102, 124, 125
Integral of measure, total, 44
Integral spin, 19
Integrals, evaluating some, 87
Integration of equations of motion, 41
Interacting masses, 3
Interaction, 28
  distances between the particles, 15
  electromagnetic, 16
  gravitational, 16
  laws of, 3
  magnetic, 30
  strong, 16
  weight of, 24
Interaction potential, two-body, 12
Interactions,
  equal weight for, 24
  equally probable, 24
  three different effective, 56
Intermediate root, 101, 102
Internal energy of complex system, 16
Internal energy of isolated system, 16
Internal state of excitation of system, 16
Interparticle coordinate vectors, 15
Interval,
  averaging, 132, 133, 144
  fixed, 138
  infinitesimal, 31
  measuring, 132
  unit, 7
Interval between adjacent roots, 46
Interval of length, 31
Interwoven runs of levels, 38
Invariance, threefold way of, 21
Invariance and equal probability, 27-28
Invariance of ensemble, 28
Invariance of Hamiltonian, 15, 19, 21

Invariant combinations of spin, 15
Invariant function, 24, 79
Invariant group measure, 30
Invariant measure, 72
Invariants of matrix, 85
Inversion transformation,
  operator of, 16
  spatial, 16, 17
Irregularity, 2
Isolated system, internal energy of, 16
Isolated system of particles, 15
Isotropy of space-time, 21

J

J (angular momentum), 4
Jacobian, 44, 78
Jacobian of the transformation, 78-79
James, A. T., 86, 88, 94, 95, 99, 158, 159, 160
Joint characteristic function, 99, 100
Joint densities of eigenvalues, 88
Joint density, 97, 98, 99, 100, 102, 103, 109, 118
Joint density of roots, 97, 100, 103, 104, 108
Joint distribution, 118
Joint distribution function, 23
Joint distribution of ratios, 109
Joint distribution of roots, 88, 95, 100, 103
Joint-probability densities, 140
Joint probability integral, 102
Joos, G., 84, 158
Jouris, G. M., 102, 107, 161

K

Kabe, D. G., 99, 160
Kahn, P. B., 49, 50, 155, 156
Kendall, M. G., 149, 166
Kernel function, 50
Khatri, C. G. 88, 92, 102, 114, 159, 160, 161, 163
Khinchin, A. L., 130, 146, 165
Kinetic energy, 16
Kisslinger, L. S., 2, 151
k-level correlation functions, 140
Koopman, B. O., 4, 152
k-point function, 142, 145
Kramers, H. A., 19, 70, 153
Kramers degeneracy, 19-20, 70
Krishnaiah, P. R., 85, 86, 88, 94, 98, 100, 102, 103, 104, 107, 108, 114, 118, 124, 158, 159, 160, 161, 162, 163
Kronecker product, 122

L

Landau, L. D., 15, 47, 153, 155
Lane, A. M., 50, 66, 156
Laplace transformation, 105, 106
Large integer *n*, 10
Large set of theoretical replicas, 41
Large shell-model calculations, 39
Large spacings, 38, 47
Large spectra, 56
Largest root, 101, 102
Latent roots, 97, 98, 99, 100, 119
Lauritsen, T., 4, 5, 152
Law,
  decay, 12
  Maxwell-Boltzmann, 146
  Poisson, 32
  Porter-Thomas, 61
  probability, 7
  semicircle, 26, 85
  spacing, 33, 42
  Wigner's, 32
Law of conservation of total linear momentum, 16
Law of motion, Newton's, 4
Laws, unknown, 3
Laws of interaction, 3
Leadbetter, M. R., 130, 165
Lebowitz, J. L., 130, 165
Lee, J. C., 110, 114, 122, 162, 163, 164
Leff, H. S., 50, 156
Legendre polynomials, 50
Lehmer, D. N., 10, 11, 153
Length interval, 31
Level,
  energy, 31
  neighborhood of lowest energy, 26
Level densities, 83

Level density, 45, 49, 50, 86, 135, 136
mean, 49
Level deviation, 54
Level distribution of nucleus, 25
Level distributions, energy, 24, 25
Level repulsion, 141
Level repulsion in spectra, 30, 31-39, 41,136
Level spacking, average, 29
Level spacing distribution, 60
Level spacings, 75, 83
energy, 44
Level structure, 2, 30
Level-to-level deviations, 56
Level-to-level fluctuations, 56
Level widths, 75
Levels, 7, 33, 37, 138, 139
actual density of, 54
alternate, 50
average width of, 7
completely random sequence, 38
density of, 7, 26, 32
discrete, 12, 42
discrete energy, 42
energy, 1, 2, 3, 4, 7, 10, 33, 70, 87, 137, 145, 150
equidistant, 7
fraction of, 7
functions of energy, 24
individual, 2
neutron-resonance, 143
nucleus with infinite number of energy, 24
number of, 24, 136
repulsion of, 46-47
repulsion of energy, 46
runs of, 33
sequence of, 49, 50
sequence of energy, 50
series of, 47
spectra of energy, 38
splitting of, 29
statistical theory of, 2-3, 16, 42, 81
theoretical calculation of energy, 13
total number of, 24, 53
transition probability, 4
widths of energy, 12, 57-58
Levels of parity, 10
Lifshitz, E. M., 15, 153
Liggett, W. S., Jr., 88, 159
Light nuclei, 41
Likelihood ratio statistics, 88, 109, 110, 112, 113, 114
Likelihood ratio tests, 110
Limiting distribution, 54
Limiting operations, 10
Line element, 76
Linear dependence of angles on energy levels, 24
Linear momentum, 16
Linear operator, 3, 16
Linear repulsion, 32
Liouville theorem, 33
Local smoothing, 54
Local spacing, mean, 32, 56, 137
Local spacing units, 51
Local spacings, 56
Local spectral averaging, 132
Local uniformity, 51
Locally uniform spectrum, 51
Long-range order properties, 49
Long-range rigidity, 51
Low-energy region, 4
Low-energy neutron, 7
Low energy spectra, repulsion in, 39-42
Low excited states, 13
Low-lying energy levels, 7
Low-lying excited states, 2
Low-lying part of spectrum, 42
Lowest energy level, 7, 26
Lowest state, density in neighborhood of, 26

M

Magic nuclei, 41
Magnetic dipole, 12
Magnetic field, 29
Magnetic interaction, 30
Magnetic moment, 58
Main group of nuclei, 41, 42
Major probability, 49, 50
Malin, S., 77, 157
Mantzouranis, G., 150, 167
Marginal distributions, 88, 109
Marginal distributions of few roots, 100-104

Mass number, 41
Masses, definite, 3
Mathematical theory of random matrices, 8
Mathemathically basis, significant, 54
Mathematics of number sequences, 10
Matrices, 3, 24
  2 × 2, 45, 46
  commuting Hermitian quaternion-real, 70
  complex, 20
  complex random, 88
  distributions of some random, 94-100
  ensemble of Hermitian, 3, 59
  ergodic properties of random, 8
  finite and infinite, 86
  Hermitian, 77
  infinite, 3
  mathematical theory of random, 8
  multivariate beta, 88, 96
  practically all self-adjoint, 84
  random, 8, 26, 81, 86, 88, 95, 100, 102, 103, 104
  real symmetric, 19, 28, 84
  self-adjoint, 84
  self-dual quaternion, 72
  set of, 67
  small, 42
  space of, 26
  space of unitary self-dual quaternion, 72
  spectral density, 122
  spin, 18, 20
  symmetric, 19, 28, 127
  unitary, 30
  unitary self-dual quaternion, 72
  vast majority of all self-adjoint, 84
  very high dimensionality, 3
  Wishart, 88
Matrices and quaternions, 69
Matrices of high order, random, 47
Matrix,
  arbitrary symplectic, 72
  banded diagonal, 19
  central complex bivariate beta, 107
  central complex bivariate *F*, 107
  central complex Gaussian, 96
  central complex multivariate, 96
  central complex multivariate beta, 96, 103, 109, 110
[Matrix]
  central complex multivariate *F*, 96, 120
  central complex Wishart, 96, 99, 102, 103, 109, 116, 119, 123
  central multivariate beta, 99
  central Wishart, 102
  characteristic values of, 46
  complex, 86
  complex canonical correlation, 103
  complex Gaussian ensemble, 103
  complex multivariate beta, 103
  complex Wishart, 87, 88, 96, 97, 99, 103, 116, 118, 119, 123, 125
  covariance, 95, 99, 110, 114, 115, 118, 122, 125
  diagonal, 77
  double root of, 46
  function of Hamiltonian, 24
  Gaussian, 98
  Hamiltonian, 18-19, 24, 25, 39, 76
  Hermitian, 3, 30, 77, 94, 127
  Hermitian quaternion-real, 70
  Hermitian random, 96
  infinitesimal Hermitian, 30
  invariants of, 85
  multivariate beta, 88, 96, 102
  noncentral complex Wishart, 123
  orthogonal, 27, 78
  population parameter, 100
  quaternion-real self-dual infinitesimal, 72
  random, 26
  real orthogonal, 24, 27, 72
  real symmetric, 70, 76
  real symmetric infinitesimal, 27
  real symplectic, 70
  roots of, 46
  S, 28, 130
  self-dual, 69
  shell-model, 42
  skew-symmetric unitary, 19
  spectral density, 87, 115, 125
  symmetric, 46
  symmetric unitary, 26
  symplectic, 70, 72
  transformation, 28
  transpose of, 19
  unit, 20
  unitary, 24, 26, 27, 28, 30, 67, 69, 72, 73

[Matrix]
unitary self-dual quaternion, 70
unitary symmetric, 72
Wishart, 102
Matrix element distribution, 43, 44, 136
Matrix elements, 58, 59, 79, 86
components of, 85
statistics of, 58
Matrix models, 86
Maxwell-Boltzmann law, 146
McKeown, M., 5, 152
Mean density, 11
Mean distance, 12
Mean level density, 49
Mean local spacing, 32, 56, 137
Mean spacing, 45, 50, 54
Mean-square error, 142
Mean vector, 95, 110, 122
Measurable physical quantity, 16
Measure, 27, 43, 44, 73, 76, 77, 78, 84, 85, 131
higher-than-two-point, 145
invariant, 72
invariant group, 30
one-point, 130, 135-139
proper, 4
total integral of, 44
two-point, 130
Measures, 143
ergodic, 135
fluctuations, 129, 130, 139, 142
one-point, 130, 135-139
Measuring interval, 132
Mechanics,
classical, 3
classical statistical, 33, 41, 130, 132
quantum, 1, 3
statistical, 2, 3, 42
Mehta, M. L., 47, 49, 50, 51, 103, 135, 138, 141, 143, 155, 156, 161, 166
Mello, P. A., 30, 50, 60, 129, 146, 150, 155, 166, 167
Messiah, A., 16, 153
Meteorological forecasts, 88
Metric, 76
Mihelich, J. W., 5, 6, 152
Mixed sequences, 33, 38
Mixed spectra, 145
Mixed spectrum, 145
Model,
Brownian motion, 50
independent particle, 85, 86
optical, 83
Model state, 42
Models, matrix, 86
Molecular physics, 30
Moment,
dipole, 58
magnetic, 58
Moment variances, 141
Momenta, 3, 4
angular, 4, 5, 18
Momenta, generalized, 3
Moments, 104, 111, 112, 113, 114, 135, 136, 138, 139, 146, 148
Moments in atomic physics, 59
Moments of elementary symmetric functions, 104-108
Momentum, 16
angula, 47
conservation of angular, 16
linear, 16
total, 16
total angular, 17, 18, 19, 21
Mon, K. K., 136, 166
Monte-Carlo calculations of quantities, 49, 65
Motion,
Brownian, 50, 86
constants of, 4, 10, 16
equations of, 41, 84
Newton's law of, 4
Multiple coherence, 99
Multiple time series, 87, 88, 122, 125
Multiplication table, 68
Multipole external crystalline electric field, 19
Multitude of systems, 4
Multivariate beta matrices, 88, 96, 102
Multivariate distributions, 8, 87-114
applications of, 115-127
Multivariate matrix, central complex, 96
Münster, A., 130, 165

N

Nagarsenker, B. N., 110, 114, 162, 163
Narrow resonances, 42

Nature, regularity of statistical, 38
Nature of system, 3
Nearest-neighbor spacing, 45, 49
Nearest-neighbor spacing distribution, 12, 31, 38, 45, 46, 47, 50, 56
Nearest-neighbor spacings, 42, 45
  histogram of, 38
Neighborhood, 73
  infinitesimal, 72
Neighborhood of lowest energy level, 26
Nested hypotheses, 123
Neutron,
  low-energy, 7
  slow, 51
Neutron-capture region, 2
Neutron-capture resonance, 28
Neutron emission, 59
Neutron fission radiative and reaction widths, 83
Neutron region, slow, 51
Neutron-resonance levels, 143
Neutron widths, 62, 64
Neutrons, slow, 38
Newton's law of motion, 4
Next-nearest-neighbor spacing distributions, 50
Noncentral complex Wishart matrix, 123
Nonisotropic external field, 21
Non-random distributions, 143
Nonstatistical phenomena, 62
Normalization condition, 53, 79
Normalization factor, 45
Normalized spacings, 41
Nuclear boundary, 7
Nuclear Hamiltonian, 84
Nuclear physics, 30, 87
Nuclear reactions, 150
Nuclear spectra, discrete, 41
Nuclear-table data, 56
Nuclear surface, 57
Nuclear table, 62
Nuclei, 3
  beryllium, 4
  boron, 4
  carbon, 4
  complex, 42
  excluded, 42
  heavy, 2
  light, 41
[Nuclei]
  lowest state in, 26
  main group of, 41, 42
  spectra of excluded, 42
  types of, 41
Nucleus, 2, 29, 84
  complex, 2, 3
  ^{167}Er, 33
  heavy, 51
  ^{180}Hf, 5
  level distribution of, 25
  ^{238}U, 7
  ^{239}U, 7
  ^{49}V, 33
Nucleus with infinite number of energy levels, 24
Nuisance parameters, 124, 125, 126
Number,
  angular momentum quantum, 5, 7
  mass, 41
  quantum, 2, 10
  wave, 54
Number of dimensions, odd, 16
Number of levels, 24, 53, 136
  total, 53
Number of particles, 3
Number of spacings, 37
Number sequences, 10
Number statistic, 136, 142
Number variance, 142
Numbers,
  complex, 24
  density of prime, 10, 11
  good quantum, 41
  ordered sequence of random, 47
  prime, 10, 11
  quantum, 33, 38, 47, 50, 145
  random, 12
  sequence of, 10-12
  sequence of random, 12
Numerical tests, 47

O

Odd number of dimensions, 16
Odd parity, levels of, 10
Odd-spin analog of orthogonal ensemble, 72
Odd-spin case, 72
Odd spin Hamiltonian, 70

Ohkubo, M., 143, 166
Olson, W. H., 136, 166
One-dimensional hydrogen atom, 10
One-dimensional quantum mechanical problems, 9
One-dimensional systems, 9-10
One-point function, 134
One-point measures, 130, 135-139
Operation of time inversion, 17
Operation of time reversal, 71
Operations, limiting, 10
Operator,
  antiinvariant, 58
  antilinear, 16, 17
  antiunitary, 17
  commuting, 18
  complex conjugation, 17
  energy, 7
  expectation values of, 12
  Hamiltonian, 1, 3, 12
  Hermitian, 1, 2, 7, 58
  identity, 17
  inversion transformation, 16
  linear, 3, 16
  parity, 16
  parity transformation, 16
  purely imaginary, 58
  real Hermitian, 7
  real symmetric, 7
  self-adjoint, 3, 58
  skew-symmetric, 58
  symmetry, 17
  time-inversion, 17
  time-reversal, 17, 20
  time-inversion-invariant, 58
  translation, 16
  unitary, 16, 17
Optical model, 83
Orbital part of angular momenta, 18
Order properties, long-range, 49
Ordered roots, 102
Ordered sequence of random numbers, 47
Ordinary statistical mechanics, 2, 132
Orthogonal distribution, 26
Orthogonal ensemble, 8, 21, 23, 24-25, 26, 29, 50, 72, 75, 80, 81, 142
  circular, 49, 50
  Gaussian, 49, 136, 138
  odd-spin analog of, 72
Orthogonal group, 19, 21, 78
Orthogonal matrix, 27, 78
  real, 72
Orthogonal transformation, 77
Oscillator, harmonic, 9

P

Pairs, equal, 70
Pandey, A., 30, 50, 60, 129, 130, 136, 137, 140, 141, 142, 145, 146, 150, 155, 165, 167
Parameter,
  cut-off, 56
  spacing, 56
Parameter space, 76
Parameters, nuisance, 124, 125, 126
Parameters of matrix, space of, 46
Parity, 2, 4, 10, 43, 47, 50
  even, 10
  odd, 10
Parity degenerate energy levels, 10
Parity operator, 16
Parity transformation operator, 16
Partial coherence, 99
Particle- and photon-induced reactions, 83
Particle emission, 12
Particle model, independent, 85, 86
Particles, 3, 15
  Brownian, 86
  isolated system of, 15
  number of, 3
  system of, 3, 12, 18
Particles in Brownian motion, 86
Partition, 94, 100, 106
Parzen, E., 116, 163
Pathak, P. K., 121, 164
Pauli spin matrices, 18, 20
Pearson's distribution, 108, 114
Penrose, O., 130, 165
Percentage points, 103, 107, 109, 110, 114
Petersen, J. S., 44, 155
Pevsher, M. I., 50, 156
Phase and time averaging, 42
Physical quantitity, measurable, 16
Physical system, 1, 2, 7, 15, 42
  aspects of, 8

[Physical system]
symmetry properties of, 8, 13, 15-21
Physically basis, significant, 54
Physics,
atomic, 30
molecular, 30
nuclear, 30, 87
Picket-fence spectrum, 51
Pilcher, V. E., 157
Pillai, K. C. S., 102, 107, 110, 161, 162
Planck's constant (h), 5
Points, percentage, 103, 107, 109, 110, 114
Poisson distribution, 12, 32, 33, 38, 39, 41, 47, 142
Poisson ensemble, 141, 146
Poisson ensembles, 136, 139
Poisson estimate, 142, 143
Poisson formula, 143
Poisson law, 32
Polynomial, 78, 148
zonal, 94, 95
Polynomials, 149
classical, 50
Legendre, 50
Population parameter matrix, 100
Porter, C. E., 9, 10, 11, 13, 15, 21, 24, 49, 50, 57, 59, 64, 75, 149, 153, 154, 155, 156, 157
Porter-Thomas distribution, 58-60, 62, 64, 65, 146, 147, 149
Porter-Thomas distribution and experimental data, 62-66
Porter-Thomas feature, 61
Porter-Thomas law, 61
Positions of energy levels, 10
Positive energies, 10
Potential, two-body interaction, 12
Potential scattering, 13
Potentials, 10
Practically all self-adjoint matrices, 84
Preview, 8
Preliminaries, 88-94, 115-116, 130-135
Preistley, M. B., 88, 159
Prime number sequence, 11
Prime number theorem, 10
Prime numbers, 10, 11
density of, 11
Probabilities, transition, 4, 5
Probability, 31
conditional, 31
differential, 12, 77
equal, 27
major, 49, 50
Probability density, 32, 53
Probability distribution, uniform, 24
Probability for spacings, 7, 46
Probability integral, 88, 100, 102, 124, 125
Probability law, 7
Problems, one-dimensional quantum mechanical, 9
Process, averaging, 4
Processes, theory of random, 4
Product, scalar, 16
Proper Hamiltonian, 4
Proper measure, 4
Proper smoothing, 54
Properties,
long-range order, 49
symmetry, 15-16, 76
Properties of physical systems, symmetry, 8, 13, 15-21
Proton resonance, 66, 132
Pulses from radioactive target, 38
Pure sequence, 33, 38, 39, 56
Purely imaginary operator, 58

Q

Quadrupole, electric, 12
Quantal spectra, 11-12
Quantities, Monte-Carlo calculation of, 49
Quantity, measurable physical, 16
Quantum mechanics, 1, 3
Quantum mechanical problems, one-dimensional, 9
Quantum number, 2, 10
angular momentum, 5, 7
Quantum numbers, 33, 38, 47, 50, 145
good, 41
Quantum spectroscopy, 12
Quantum states, system with, 24
Quantum system, 2
Quasiergodic theorem, 84
Quaternion,
conjugate, 68

[Quaternion,]
  Hermitian conjugate, 69
  real, 68, 69
Quaternion algebra, 69, 70-72
Quaternion elements, 69
Quaternion-real, 21, 69
Quaternion-real self-dual infinitesimal matrix, 72
Quaternion structure, 20-21
Quaternions, 8, 66, 67, 68-69
Quaternions and matrices, 69
Quaternions and symplectic group, 67-73

R

Radiative source, 11
Radioactive target, pulses from, 38
Rainwater, J., 2, 44, 152, 155
Random aspects, 51
Random Hamiltonians, 86
Random matrices, 8, 26, 81, 86, 88, 95, 100, 102, 103, 104, 130
  distributions of some, 94-100
  ergodic properties of, 8, 42, 129-150
Random matrices and energy levels, 83-84
Random matrices of high order, 47
Random mutually perpendicular vectors, 45
Random numbers, 12
  ordered sequence of, 47
  sequence of, 12
Random processes, theory of, 4
Random sequence, 11, 31, 38
Random sequence levels, completely, 38
Random spectra, 39
Random variables, 96
  Gaussian, 23, 32, 62, 135
Random vector, 115
Range,
  energy, 25
  infinite, 24
Range of angles, 24
Range of major probability, 49, 50
Rao, C. R., 123, 164
Rate, transition, 59
Rate of change of distributions with dimension, 49
Ratios of roots, distributions of, 108-109
Rayleigh distribution, 32
Reaction theories, resonance, 83
Real Hamiltonian, 19, 23
Real Hermitian operators, 7
Real orthogonal matrix, 24, 27, 72
Real quaternion, 68, 69
Real roots of algebraic equation, 25
Real symmetric Hamiltonian, 19, 23
Real symmetric infinitesimal matrix, 27
Real symmetric matrices, 19, 28, 84
Real symmetric matrix, 70, 76
  characteristic values of, 46
Real symmetric matrix Hamiltonian, 23
Real symmetric operators, 7
Reality of Hamiltonian, 24
Recent studies, 85-86
Region, higher-energy, 4
Regularity of statistical nature, 38
  low-energy, 4
Relation to Kramers degeneracy, 19-20
Relative error, 135, 145
Remark, 50
Replicas, large set of theoretical, 41
Representation, 79
  coordinate, 17, 18
Representation of states, 28
Representation of system, 24
Repulsion, 32, 33
  level, 30, 31-39, 41, 136, 141
  linear, 32
Repulsion in low energy spectra, 39-42
Repulsion in spectra, level, 30, 31-39, 41
Repulsion of energy levels, 46
Repulsion of levels, 46-47
Residual level-to-level fluctuations, 56
Resonance,
  narrow, 42
  neutron capture, 28
  proton, 66, 132
  slow-neutron, 33, 41
Resonance reaction theories, 83
Reversal, time-, 16-18, 21, 69

Richert, J., 150, 167
Riemannian space, 76
Rigid spectrum, 38, 51
Rigidity, 38, 51
  long-range, 51
  spectral, 50-56, 142
rms average, 51
rms deviation, 51, 56
Role of the Hamiltonian, 84
Root,
  intermediate, 101, 102
  largest, 101, 102
  smallest, 101, 102
Roots, 2, 88, 97, 98, 102, 114, 120
  adjacent, 46
  distribution of, 86, 102, 103, 109
  distributions of ratios of, 108-109
  extreme, 102
  joint density of, 97, 100, 103, 104
  joint distributions of, 88, 95, 100
  latent, 97, 98, 99, 100, 119
  marginal distribution of few, 100-104
  ordered, 102
  unordered, 103
Roots of matrix, 46, 109
Rosen, J. L., 2, 152
Rosenzweig, N., 16, 24, 50, 59, 75, 153, 154, 157
Rotation,
  angles of, 77
Rotational band, 5
Rotational nuclei, 41
Rotations, 16, 19
Roy, S. N., 86, 118, 158, 163
Running average, 54
Runs of levels, 33
  interwoven, 38

S

Sample, 142
Scalar, 69, 70, 71
Scalar product, 16
Scattering potential, 13
Scharff-Goldhaber, G., 5, 152
Schiff, L. I., 1, 151
Schrödinger equation, 1, 2, 11, 12, 17, 19, 77
Schuurmann, F. J., 102, 103, 107, 108, 109, 120, 161, 162
Scott, J. M., 57, 157
Secular behavior, 50, 56
Secular variation, 51, 64, 65, 141
Segment of spectrum, 38, 132
Self-adjoint matrices, 84
Self-adjoint operator, 3, 58
Self-dual matrix, 69
Semicircle distribution, 25-26, 136
Semicircle law, 26, 85
Semiellipse distribution, 25, 26
Separation of fluctuations, 50
Sequence,
  pure, 33, 38, 39
  random, 11, 31, 38
Sequence of energy levels, 50
Sequence of levels, 49, 50
Sequence of numbers, 10-12
Sequence of prime numbers, 10, 11
Sequence of random numbers, ordered, 47
Sequence of random numbers, 12
Sequences,
  energy level, 50
  fluctuations for pure, 56
  mixed, 33, 38
Series of levels, 47
Set, complementary, 92, 93
Set of matrices, 67
Set of theoretical replicas, 41
Set of values, 54
Shell-model calculation, 33, 38, 39, 51, 60, 64, 65
Shell-model case, 61
Shell-model matrix, 42
Shell-model spectrum, 51, 56
Shell-model states, 65
Signal detection, 88
Similarity transformation, 18
Simultaneous eigenstates, 16
Single-eigenvalue distribution, 45, 48
Single spectrum, 41
Singularity, delta-function, 135, 140, 141
Skew-symmetric operator, 58
Skew-symmetric tensor, 68
Skew-symmetric unitary matrix, 19
Slow-neutron data, 56
Slow-neutron region, 51

Slow-neutron resonance, 33, 41
Slow neutrons, 38
Slutsky, E. E., 134, 165
Slutsky's theorem, 134
Small matrices, 42
Small spacings, 38, 47
Smallest root, 101, 102
S-matrix, 28, 130
S-matrix ensemble, 130, 150
S-matrix ensembles, ergodicity of, 150
S-matrix theory, 150
Smoothed spectrum, 51, 56
Smoothed strength, 61
Smoothing, 149
  local, 54
  proper, 54
  spectral, 136
Smorodinsky, Ya., 47, 155
Solutions, even, 10
Some random matrices, distributions of, 94-100
Sorensen, R. A., 2, 151
Source, decaying radiative, 11, 12
Space,
  basic, 28
  compact, 26
  dimension of, 76
  dimensionality of, 53
  high dimensionality, 59
  Hilbert, 3, 59, 84
  parameter, 46, 76
  Riemannian, 76
  statistical, 146
  volume of, 30
Space of parameters of matrix, 46, 76
Space of symmetric unitary matrices, 27
Space of unitary matrices, 30
Space of unitary self-dual quaternion matrices, 72, 73
Space-time
  homogeneity of, 21
  isotropy of, 21
Space-time translation, 17
Spacing, 12, 45, 56
  average, 33, 37, 41, 46, 54
  mean, 45, 50, 54
  mean local, 32, 56, 137
  nearest-neighbor, 45, 49
  spectral-averaged, 131, 137
Spacing distribution, 38, 41, 46, 47
  nearest-neighbor, 12, 31, 38, 45, 46, 47, 50, 56
Spacing distributions, 45, 49, 50
  higher-order, 49
Spacing histogram, 39, 41
Spacing law, 33, 42
Spacing parameter, 56
Spacing variances, 131
Spacings, 41
  absence of small, 47
  distribution of, 87
  distribution of nearest-neighbor, 42
  energy levels, 10, 44
  histogram of nearest-neighbor, 38
  large, 38, 47
  level, 75, 83
  local, 56
  normalized, 41
  number of, 37
  probability for, 7, 46
  small, 38, 47
  spectral-averaged, 137
Spatial axes, 16
Spatial inversion, 16, 17
Spatial inversion transformation, 16
Specific system, 4
Specification of the distribution function, 79-80
Spectra, 9, 16, 33, 38, 39, 130, 132
  complex, 68
  discrete nuclear, 41
  energy levels, 38
  large, 56
  level repulsion in, 30, 31-39
  mixed, 145
  quantal, 11-12
  random, 39
  unfolded, 140
Spectra of excluded nuclei, 42
Spectral average, 135, 141, 143, 146, 147
Spectral-averaged k-th order spacing, 137
Spectral-averaged local spacing, 131
Spectral-averaged spacings, 137
Spectral averaging, 130, 131, 132, 133, 135, 136, 137, 142
Spectral density function, 149
Spectral density matrices, 122

Spectral density matrix, 87, 115, 125
Spectral moments, 149
Spectral rigidity, 50-56, 142
Spectral smoothing, 136
Spectrally-smoothed density function, 149
Spectroscopy, quantum, 12
Spectroscopy of highly excited complex systems, 8, 9
Spectrum, 33, 51, 54, 56, 130, 142, 143
  complete, 132
  density of discrete, 51
  deviation of, 51
  entire, 39, 51, 132
  fluctuation-free, 51
  histogram of, 56
  individual, 42
  locally uniform, 51
  low-lying part of, 42
  mixed, 145
  picket-fence, 51
  rigid, 38, 51
  segment of, 38, 132
  shell-model, 51, 56
  single, 41
  smoothed, 51,56
  statistical properties of, 4
  unbounded, 53
  unfolded, 131
  uniform, 38, 137
  uniformly spaced, 51
Spectrums, 38
Spin, 2, 17, 18, 43, 50, 62
  half-integral, 19
  integral, 19
  invariant combination of interparticle coordinate vectors and, 15
Spin matrices, 18, 20
Splitting of levels, 29
Square of wave function, 7, 57
Square well, infinite, 9
Srivastava, M. S., 97, 160
Staircase function, 53
State,
  excitation of internal, 16
  final, 28, 66
  first-excited, 41
  ground, 10, 39, 41, 42, 51, 60, 66
  initial, 28, 66
[State]
  lowest, 26
  model, 42
  three-dimensional hydrogen s, 10
State function, 1
States, 129, 147
  angular momenta of, 5
  doorway, 62
  ensemble of, 3
  excited, 2, 10
  individual, 38
  low excited, 13
  representation of, 28
  shell-model, 65
  stationary, 2, 4
Stationary, 42, 130, 132, 133, 141, 150
Stationary Gaussian multivariate time series, 115
Stationary states, 2, 4
Statistic, 105
  additive, 137
  number, 136, 142
Statistical concepts, 11
Statistical error, 135
Statistical hypothesis, basic, 24
Statistical mechanics, 2, 3, 42, 130
  classical, 33, 41
  comparison with, 3-4
  ordinary, 2, 132
  theory of, 84
  usual, 3
Statistical nature, regularity of, 38
Statistical properties of,
  spectrum, 4
  symmetric operators, 7
Statistical space, 146
Statistical theory of nuclear reactions, 150
Statistical theory of energy levels, 2-3, 16, 42, 81
Statistical weight, 24, 30, 73
Statistically independent components, 24
Statistics of matrix elements, 58
Statistics of resonance parameters, 83
Strength distribution, 146
Strength distributions, 129
  behavior of, 146-150
Strength ergodicity, 146

T

Testing hypotheses, 116-122
Theorem,
  central limit, 96
  Diananda's, 135
  entropy, 84
  equipartition, 84
  Liouville, 33
  prime number, 10
  quasiergodic, 84
  Slutsky's, 134
Theoretical calculation of energy levels, 13
Theoretical replicas, 41
Theory,
  information, 86
  S-matrix, 150
Theory of levels, statistics, 2-3, 16, 81
Theory of random processes, 4
Theory of statistical mechanics, 84
Thermodynamical systems, 42
Thomas, R. G., 57, 64, 149, 157
Three-dimensional hydrogen *s* state, 10
Threefold way of invariance, 21
Time,
  analog of, 42
  direction of, 16
  functions of, 3, 4
Time average of function, 41
Time-averaged properties, 84
Time averages, 3
Time axis, 42
Time-inversion-invariant operator, 58
Time inversion operator, 17
Time inversion symmetry, 7
Time-reversal, 16-18, 21, 69
  Hamiltonian without, 30
Time-reversal invariant Hamiltonian, 18, 19, 21
Time-reversal invariant system, 24
Time-reversal operation, 71
Time-reversal operator, 17, 20
Time-reversal symmetry, systems without, 29-30
Time symmetry, 7
Time translation, 15
Tinkham, M., 19, 70, 153
Tong, H., 88, 159
Total angular momentum, 17, 18, 19, 21
Total energy, 16
Total energy of excitation, 25
Total ignorance, 28
Total integral of measure, 44
Total linear momentum, 16
Total momentum, 16
Total number of levels, 24, 53
Total volume, 27, 73
Traces distributions, 88, 107
Transformation, 78
  canonical, 16, 18, 19, 20, 21
  improper, 16
  inversion, 16
  Jacobian of, 78-79
  Laplace, 105, 106
  orthogonal, 77
  parity, 16
  similarity, 18
  symplectic, 20
  unitary basis, 19
Transformation matrix, 28
Transformation operator, parity, 16
Transformations,
  symplectic group of, 20
  unitary, 85
Transition probabilities, 4, 5
Transition rate, 59
Transition strengths, 64, 130, 135
Transitions, 64, 65, 129
Translation,
  space-time, 17
  time, 15
Translation operator, 16
Transposition, 69
Trees, R. E., 50, 156
Turin, G. L., 100, 160
Turlay, R., 7, 153
Two-body interaction potential, 12
Two-body nature of Hamiltonian, 33
Two-point function, 139, 143, 149
Two-point measures, 130
Two-state functions, 16

U

Ullah, N., 59, 157
Unbounded spectrum, 53
Unfolded spectra, 140
Unfolded spectrum, 131
Unfolding, 132, 133

Uniform distribution, 26
Uniform ensemble, 30
Uniform probability distribution, 24
Uniform spectrum, 38, 137
Uniformly spaced spectrum, 51
Unit circle, 24, 26, 70, 72
Unit interval, 7
Unit matrix, 20
Unitary basis transformation, 19
Unitary ensemble, 8, 28, 29-42, 50, 80-81, 142
Unitary ensemble, circular, 50
Unitary group, 18, 21, 30
Unitary matrices, 30
  space of, 30
Unitary matrix, 24, 26, 27, 28, 30, 67, 69, 72, 73
Unitary operator, 16, 17
Unitary self-dual quaternion matrix, 70
  space of, 72, 73
Unitary skew-symmetric matrix, 19
Unitary symmetric matrix, 72
Unitary transformation, canonical, 20
Unitary transformations, 85
Units, local spacing, 51
Unknown laws, 3
Unordered roots, 103
Uppuluri, V. R. R., 136, 166
Uranium, 7

V

Values,
  characteristic, 2, 4, 59, 86
  expectation, 44, 57, 59, 75
  initial, 4
  set of, 54
Values of operators, expectation, 12
van der Pol, B., 10, 153
Variables, 77
  discrete, 31
  eigenvalue, 78
  eigenvector, 78
  Gaussian random, 23, 32, 62, 135
  Hamiltonian matrix element, 78
  random, 96
Variance, 32, 60, 61, 96, 131, 132, 133, 134, 135, 136, 138, 142, 143, 148, 149
Variate complex normal, 99
Variation, secular, 51, 64, 65
Vast majority of all self-adjoint matrices, 84
Vector,
  basis, 146
  mean, 95, 110, 122
  random, 115
  zero mean, 115
Vectors, 59
  characteristic, 2
  interparticle coordinate, 15
  random mutually perpendicular, 45
Very high dimensionality matrices, 3
Vibration of airframe structures, 88
Volume, total, 27, 73
Volume element, 27, 76, 77
Volume of space, 30
von Neumann, J., v, 4, 46, 77, 152, 155

W

Wahba, G., 110, 116, 162, 163
Waikar, V. B., 86, 98, 102, 103, 118, 124, 158, 160, 161, 164
Wave function, 7, 12, 57
  square of, 7, 57
Wave number, 54
Wavelength, 56
Wavelengths, components of, 54
Weidenmüller, H. A., 150, 167
Weight,
  equal, 24
  statistical, 24, 30, 73
Weight of interactions, 24
Weighting function, 33
Well, square, 9
Weyl, H., 20, 67, 154
Wick, G. C., 16, 153
Wide class of systems, 54
Widths, 44, 150
  average, 7
  distribution of, 7-8, 57-66
  level, 75
  neutron, 62, 64
Widths of energy levels, 12, 57-58
  fluctuation properties of, 57

Wigner, E. P., v, 3, 5, 16, 23, 25, 46, 47, 58, 59, 75, 84, 85, 98, 103, 152, 153, 154, 155, 158, 159
Wigner's asymptotic semicircle law, 49
Wigner's conjecture, 47-49
Wigner's distribution, 32, 33, 38, 39, 41
Wigner's distribution function, 47
Wigner's formula, 47
Wigner's law, 32
Wigner's semicricle distribution, 136
Wilks, S. S., 118, 164
Wilson, K., 19, 154
Wishart distribution, 85
Wishart ensemble, 26-27, 86
Wishart matrices, 88
Wishart matrix, 102
    complex, 87, 88, 96, 97, 99, 103, 116, 118, 119, 123, 125
Wong, S. S. M., 30, 50, 60, 85, 129, 155, 158
Wooding, R. A., 88, 96, 159

Y

Yaglom, A. M., 130, 134, 165
Young, D. L., 102, 161
Young, J. C., 123, 164

Z

Zero field, 30
Zero mean vector, 115
Zero means, 99
Zonal polynomial, 94, 95